THE ENERGY MACHINE OF T. HENRY MORAY

OTHER BOOKS BY MORAY B. KING:

- **TAPPING THE ZERO-POINT ENERGY**
- **QUEST FOR ZERO-POINT ENERGY**

THE ENERGY MACHINE OF T. HENRY MORAY

Zero-Point Energy & Pulsed Plasma Physics

by

Moray B. King

THE ENERGY MACHINE OF T. HENRY MORAY

Zero-Point Energy & Pulsed Plasma Physics

Adventures Unlimited Press

The Energy Machine of T. Henry Moray

ISBN 1-931882-42-8

Published by

Adventures Unlimited Press
One Adventure Place
Kempton, ILLINOIS 60946 USA

Printed in the United States of America

Published in association with
Paraclete Publishing
P.O. Box 859
Provo, UT 84603

www.adventuresunlimitedpress.com
www.adventuresunlimited.nl

Dedication

In rememberance of Edwin Gray, Stephen Marinov, Stan Meyers, Paul Brown, Eugene Mallove.

They gave their lives for their efforts to bring a new energy source to mankind.

Table of Contents

THE ENERGY MACHINE OF T. HENRY MORAY

T. Henry Moray

The "free energy" invention of T. Henry Moray is probably the most famous and well witnessed in the history of the field. The best version of the device was claimed to yield 50 kilowatts of electricity without using any known input power. Radioactive material was used to maintain plasma activity in the tubes, but was too weak to account for the output energy. The electricity exhibited a strange "cold current" characteristic where appreciable power could be guided on thin wires without heating them. Moray suffered through ruthless suppression, and the device was destroyed. T.H. Moray died in 1974, but his son, John Moray, was funded to continue the research. The story continues today with experimental results from other investigators such as Paul Brown, Paulo and Alexandra Correa, Ken Shoulders and Edwin Gray, contributing significantly to its understanding.

1. **What ever happened to the T.H. Moray device?** Thomas Henry Moray is perhaps the most famous "free energy" inventor in history mainly due to the number of witnesses of the device, scientific investigators and letters of testimony. The fact that his last name is identical to my first name manifests deep personal synchronicity, for it was only after I discovered the zero-point energy (ZPE) in the physics literature and began earnestly exploring its possible technological applications that I was given the following book:

THE SEA OF ENERGY
IN WHICH THE EARTH FLOATS

2. ***The Sea of Energy in which the Earth Floats.*** (Moray, 1960, 1978) When I saw that the author's name was identical to my unusual name, I was stunned. For me it meant that my research into zero-point energy was to apply to this device, and my purpose was to explain it so clearly that the scientific and engineering community would be able to successfully create similar energy machines.

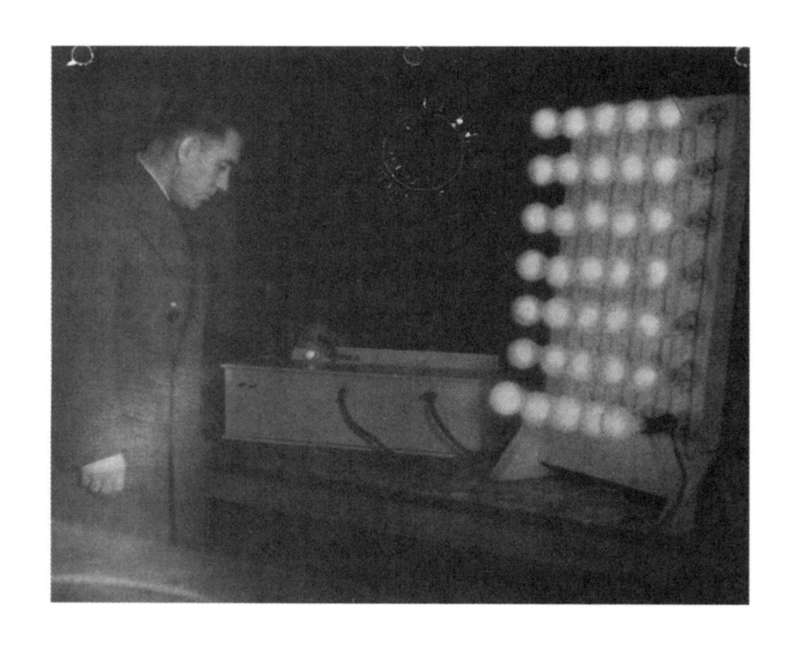

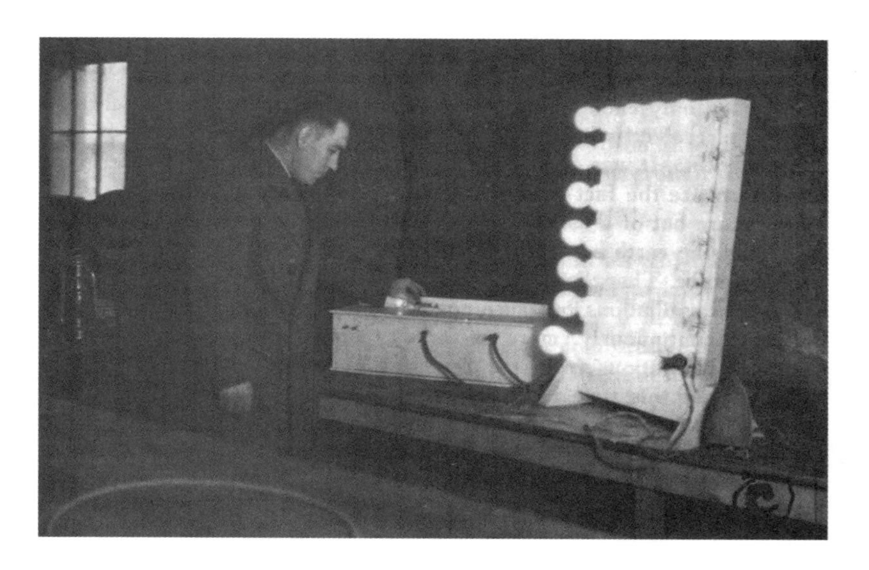

3. **T.H. Moray and bulbs.** Here was the story of an inventor and the description of an invention that output abundant electricity from an unrecognized energy source. The device could not be explained with standard classical scientific theories. Moray believed he was tapping radiant energy or ether waves impinging from space. As I read the explanations in the book, whenever Moray described his methods for tapping "radiant energy," I was amazed that I could find support for his methods in today's physics literature that seems to explain the device. The fundamental operating principle arises from a surprisingly simple hypothesis: *Abrupt, synchronous, ion surges in plasma appear to coherently activate the zero-point energy.* Although Moray's research predated ZPE theories, he empirically discovered the importance of ion oscillations.

Note dark Spot around lights
Burned in film. Not

4. Demo with antenna. Moray gave numerous demonstrations of his invention. All he would ask for in return were letters of testimony from the witnesses. As a boy Moray was a crystal radio enthusiast. The challenge in this hobby was to maximize the received signal using a good antenna and ground connection, and most importantly, a good rectifying detector. Exotic materials were often tried. While on a church mission to Sweden, Moray found the mysterious "Swedish stone," which became the main

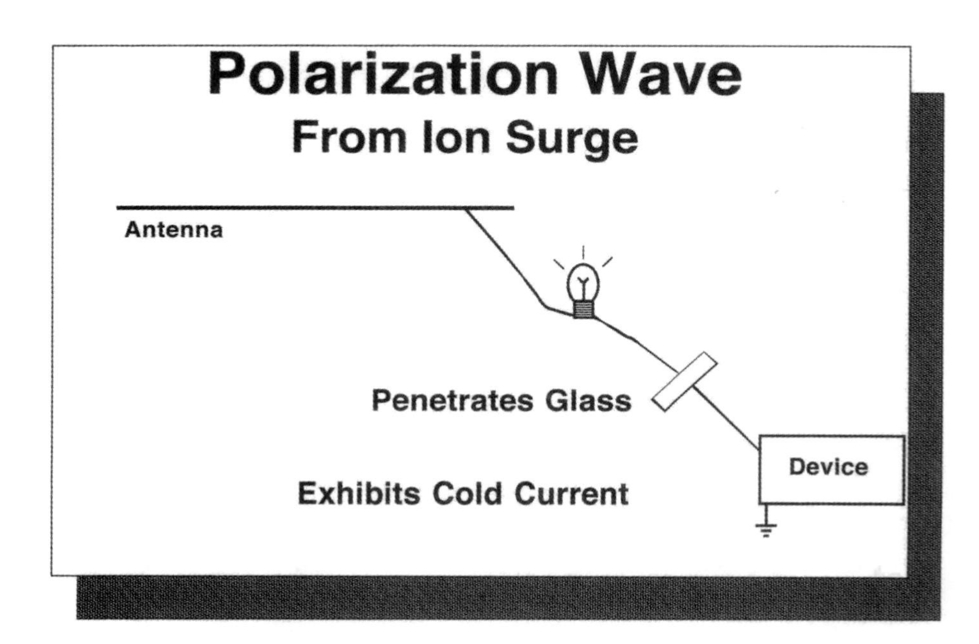
Polarization Wave
From Ion Surge
Antenna
Penetrates Glass
Device
Exhibits Cold Current

detector material for his invention. The first device could only weakly light a single bulb. Beginning in 1925 the invention was improved to a few hundred watts, and later demos typically manifested about 5000 watts. Here the device would light a bank of bulbs, power an electric heater, and drive a fan.

5. Polarization wave. The electricity coming from the device had a strange "cold" characteristic. Moray used very thin (no. 30) wire in constructing a device that output kilowatts, yet the wires did not overheat. Sometimes for a demonstration, a bulb was placed in series on the antenna lead, and it would light while the device was operating. (It would not light if the lead were directly grounded, showing the energy was not coming from the antenna-ground circuit.) Moray would cut the lead between the bulb and the running device and insert panes of window glass in the gap between the cut leads. The bulb would continue to shine. Skeptical investigators would bring their own window glass for this demo to insure it was not a trick. The hypothesis of a polarization wave, launched from abrupt ion motion, is discussed below to explain the cold current effect and the glass penetration.

R.E. Device Used in Airplane Test

6. **Final device (no antenna).** Moray eventually improved the device to where the antenna and ground were no longer needed. The device was successfully tested in an airplane, a mineshaft, and submarine. Moray claimed that a device weighing about 50 pounds would be capable of providing 50 kilowatts. Running the device in shielded environments showed that the energy source could not be standard, ambient electromagnetic fields.

Sound Pickup Device

Similar to Crystal Set, Infrasound Detector

Range of Miles

Acoustical Soliton?

Ion Polarization Waves?

7. Sound pickup device. T. H. Moray created other unusual inventions based on crystal set technology. His crystal detector could amplify a radio signal sufficiently to drive a loud speaker. Moray also invented a sound pickup device, which could be tuned to receive ordinary verbal conversions miles away. Many witnesses tried the device, and one reported hearing conversations and background sounds from the train station over four miles away. The claims are hard to believe because normal acoustical sound waves would be expected to disperse with distance. How can this device be explained? Perhaps a clue is that infrasound detectors (Vassilatos, 1991) resemble broadband crystal radio detectors (with the tuning capacitor removed), which matches Moray's energy device detection scheme. If sound near the ground were captured within infrasound surface solitary waves or if the sound launched polarization waves from residual air ions, perhaps his amplifying detector might be sensitive to them. Moray was funded in the early 1950's to improve the sound pickup device on a classified contract; the results have never been made public. Did Moray accidentally discover sound pickup properties while working with glow plasma crystal detectors?

Transmutation of Elements

Some Electrode Metal Changes to Another Element

Unusual Isotopes

Observed in Cold Fusion Cathodes

and Charge Cluster Strikes

8. Transmutation of elements. The later years of Moray's research were dedicated to investigating a peculiar anomaly that occurred on the electrodes within his plasma tubes: Some of the electrode metal would transmute to another element. It exhibited a nuclear reaction as if the nucleus absorbed a proton or emitted a beta particle. Because Moray was secretive about this research, little is known. A clue might appear in Moray's patent: There is one paragraph describing how to make a particular lead sulfide mixture to be used as electrode material. Did Moray discover how to transmute lead into gold? The anomaly regarding transmutation of elements is hard to believe, except that it has been observed over the last decade in the cathodes from cold fusion experiments (Fox, 1996), and these experiments are quite repeatable today.

Thomas Henry Moray

1892 - Born
1909 - Begins research
1911 - 1 Bulb, half power
1912 - LDS mission
1913 - Finds Swedish stone
1917 - Marries, employed as engineer
1925 - Demos begin, 100 watts
1929 - Russian interest, 600 watts
1930 - Moray Products Co.
1931 - Patent application rejected
1938 - 4500 watts
1939 - REA builds lab
1939 - Felix Frazer smashes device
1940 - Wounded in gun fight
1943 - Attempts to rebuild
1949 - Electrotherapeutic Patent
1950 - Sound device, Radio Signal Labs
1950's - 1960's Transmutation experiments
1974 - Dies

9. Timeline. The Moray story is a tragedy. The achievement of technical success was followed by business subversion, government corruption, threats, and assassination attempts. He installed bulletproof glass on his car for protection. He was wounded in a gunfight during a raid on his lab. Agents from the Rural Electrification Agency (REA) encouraged him to move to Russia. (A Congressional investigation later revealed that communists had infiltrated the REA.) In 1939 an investigator from the REA, who worked closely with Moray, took a hammer and smashed the machine's expensively crafted tubes. Moray later tried to rebuild a lesser version of the device, but burned out his detector. Because of the threats to his life and family, Moray chose not to rebuild the energy machine, but instead focused his research on another anomaly that occurred within his plasma tubes: transmutation of elements.

Project X

Glenn Foster, 1976

$600,000 - Hanscom Labs, Cambridge, Mass.
$280,000 - Eyring Research Institute
Cosray Research Institute

24 Oscillator tubes built

Swedish stone - Diatomaceous earth and quartz

Diamond press pellets with radium doping: EMP accident

J.E. Moray, E.E. Dahl Assoc, US Air Force Systems Command, #F42600-75-2212, Final Report, April 15, 1977.

10. Project X. In 1974 T.H. Moray passed away, but the research continued. His son, John Moray, was funded by an Air Force contract to rebuild the tubes of the device. Glenn Foster (recently deceased) arranged to fund a number of unusual energy projects in the mid 1970's, some of which were quite successful (e.g. the lithium battery). Foster related that on project X some of the "Swedish stone" material, which analysis showed was comprised of diatomaceous earth doped with a weakly radioactive mixture, was subjected to extreme pressure in a diamond press. An electromagnetic pulse (EMP) resulted which blew out circuit breakers and damaged a power line transformer across the street (Perreault, 1999). Over twenty oscillator tubes were built on the project, but an energy system was not constructed. There were negotiations for a follow up project, but it never happened. Today the notes from years of research and some equipment are stored with the Moray family in Canada.

Big Mysteries

1. Excess Energy
2. Cold Current
3. Sound Device
4. Transmutation

11. Big mysteries. There are four big mysteries associated with Moray's research: 1) What is the energy source driving the device? 2) How can thin wires conduct "cold currents" that penetrate glass? 3) How can a device pick up normal street conversations from miles away? 4) How can element transmutation occur at low energies? All of these anomalies seem to center around surging or oscillating ions in plasma, a theme that Moray emphasized throughout his book.

Significant Inventors

Paul Brown:
Nuclear Battery

Paulo and Alexandra Correa :
Pulsed Anomalous Glow Discharge Tube

Edwin Gray:
Electrical Conversion Switching Tube

Ken Shoulders:
High Density Charge Clusters

12. Significant inventors. Support for explaining the Moray discoveries comes from examining the experimental work and patents of four significant inventors, who appear to have developed related technologies. These include Paul Brown's nuclear battery, Paulo and Alexandra Correa's pulsed glow plasma discharge tube, Edwin Gray's pulsed plasma tube, and Ken Shoulder's high-density charge clusters.

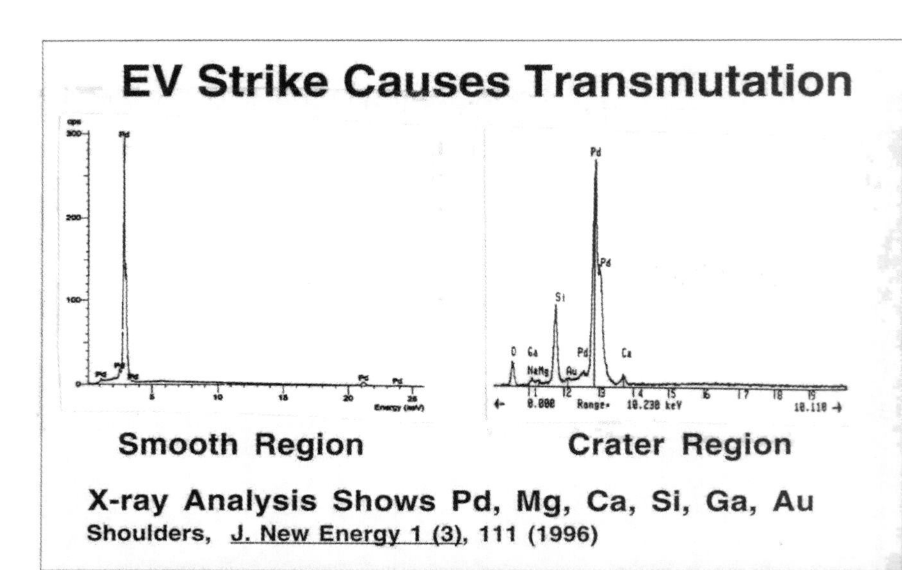
EV Strike Causes Transmutation
Smooth Region
Crater Region
X-ray Analysis Shows Pd, Mg, Ca, Si, Ga, Au
Shoulders, J. New Energy 1 (3), 111 (1996)

13. Transmutation from EV strike. Ken Shoulders (1991, 1996) has demonstrated perhaps the simplest experiment to manifest element transmutation. Shoulders discovered how to launch a coherent plasma form that appears to be a cluster of charges predominantly of one polarity, which he named "electrum validum" (EV). A single strike onto an aluminum plate from one high density charge cluster, can result in transmutation of aluminum nuclei as exhibited by a scanning electron microscope (SEM) analysis of the crater region where the EV hit. Moreover, the resulting transmuted isotopes are unusual and rarely found in nature.

Pure EV Launcher

(Cross Section of Cylinder)

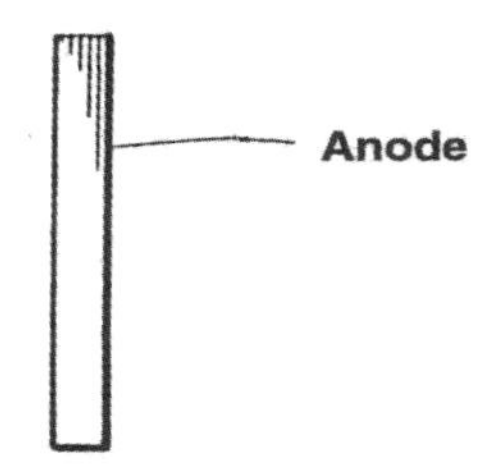

Shoulders, U.S. Patent # 5,018,180 (1991)

14. EV launcher. Charge clusters are readily launched from a liquid metal tipped electrode. They are typically about a micron in size, and exhibit a net charge of about 10^{11} electrons and can carry thousands of ions. They exhibit anomalous excessive energy that Shoulders suggests comes from the zero-point energy. He has made larger (centimeter size) versions, but the EMP blast that results when they strike a conductor damages electronic equipment, and thus makes them too dangerous to study. The patent clearly explains how to create charge clusters, and offers the scientific community a true energetic anomaly that is readily repeatable.

Liquid Metal Protuberance

Glow Plasma

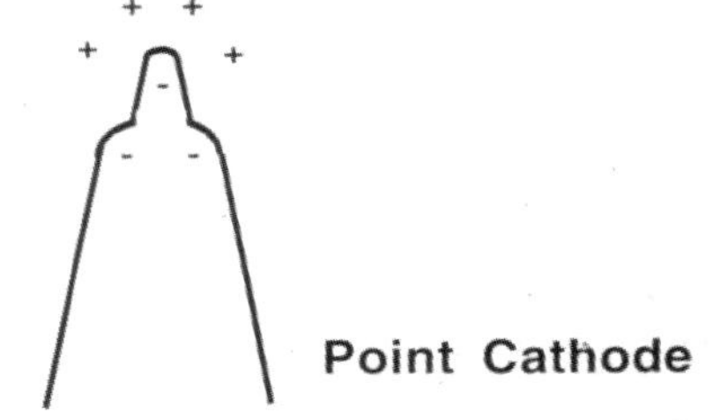

Mesyats, Proc. 17th Int. Sym. on Discharges and Electrical Insulation in Vacuum, 720 (1996)

15. Liquid metal protuberance. The charge cluster arises from perfectly symmetrical boundary conditions. Just before the emission a microscopic liquid metal stalk protrudes from the end of the pointed electrode (Mesyats, 1996). Polarized corona surrounds the stalk with the ions attracted toward the tip. The state exhibits perfectly symmetrical boundary conditions.

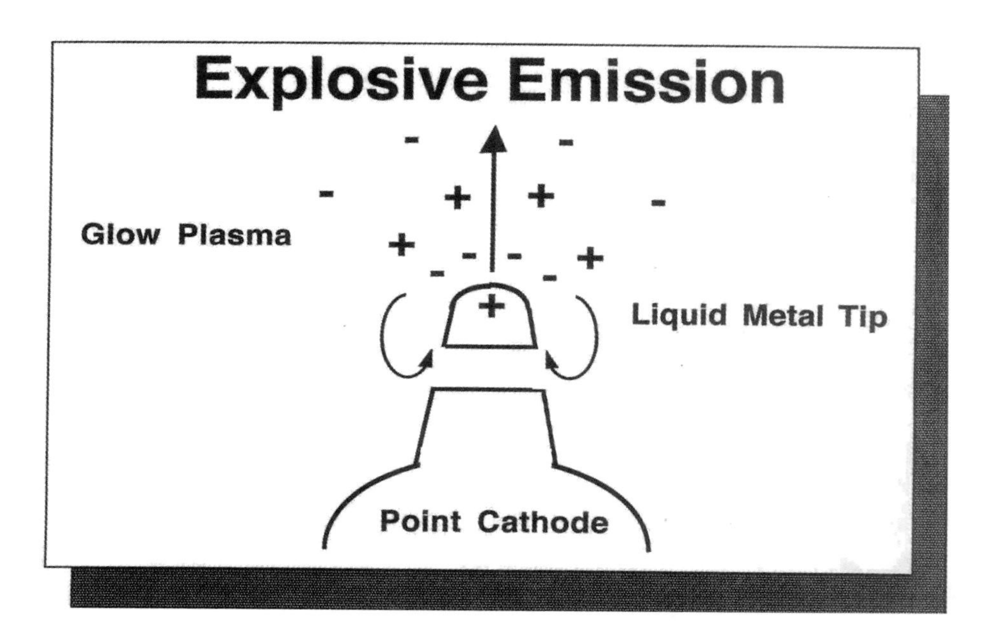
Explosive Emission
Glow Plasma
Liquid Metal Tip
Point Cathode

16. Explosive emission. The tip of the liquid metal stalk explodes off creating an abrupt compression event with the ions in the surrounding corona. We will examine similar ion compression examples to support the proposal that such an event could induce coherent ZPE coupling with the ions participating in the impulse. The perfect symmetrical geometry from the liquid metal tip guides the plasma into forming a vortex ring filament.

Helical Flow in Plasmoid Vortex Ring Filament

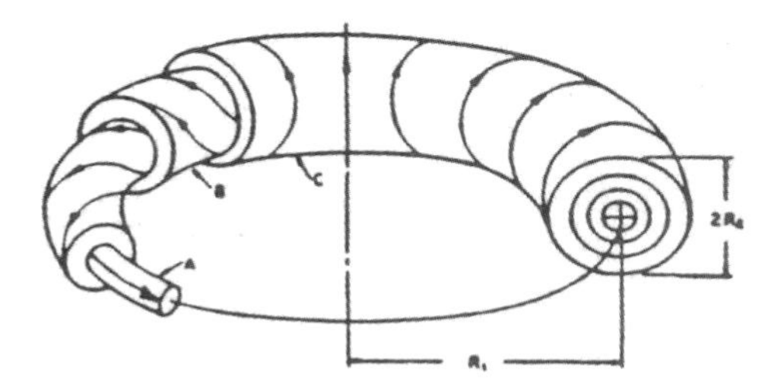

Force Free Vortex Yields Natural Stability

Alex, Radar, Fusion Tech. 27, 271 (1995)

17. Vortex ring model. Charge clusters appear to be a miniature form of ball lightning, which many investigators suggest gains stability via a vortex ring charge circulation. A like geometry might be archetypal for modeling charge and pair production arising from the underlying energetic vacuum fluctuations. Charge clusters typically occur in discharge events, and their excessive energy would likely contribute to the plasma activity occurring in Moray's tubes.

Launching Charge Cluster

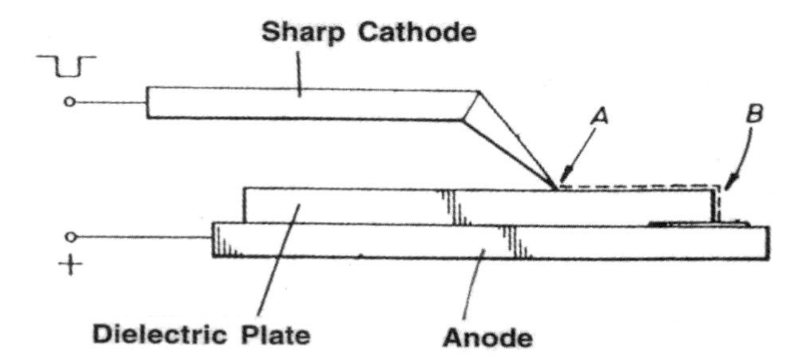

Shoulders, U.S. Patent # 5,018,180 (1991)

18. **Point contact EV launcher.** Point contact electrodes touching a dielectric surface can launch a charge cluster onto the surface. Corona around the tip guides the explosive emission to form the EV. Moray utilized a pointed electrode in contact with surface glow plasma on his crystal detector, which could induce this type of activity at a low trigger voltage. Experiments with point contact discharge into surface glow plasma might manifest excessively energetic events.

Energy Source

1. Radiant energy
2. Nuclear energy
3. Zero-point energy
4. Synergistic combination

19. Energy sources. To explain the invention researchers have hypothesized three primary sources of energy. Moray believed he was tapping radiant energy (ether waves) propagating from space. Many (Brown, 1987, Moreland, 1997, Pereault, 1999) have suggested nuclear energy is the primary source since it is well known that Moray mixed radioactive material with his detector, cathodes, and dielectrics in his tubes. But was the radioactivity powerful enough to provide kilowatts? Zero-point energy is potentially powerful enough, but the amount accessible is highly controversial in the physics community. Of course, any synergistic combination of energy sources should also be considered.

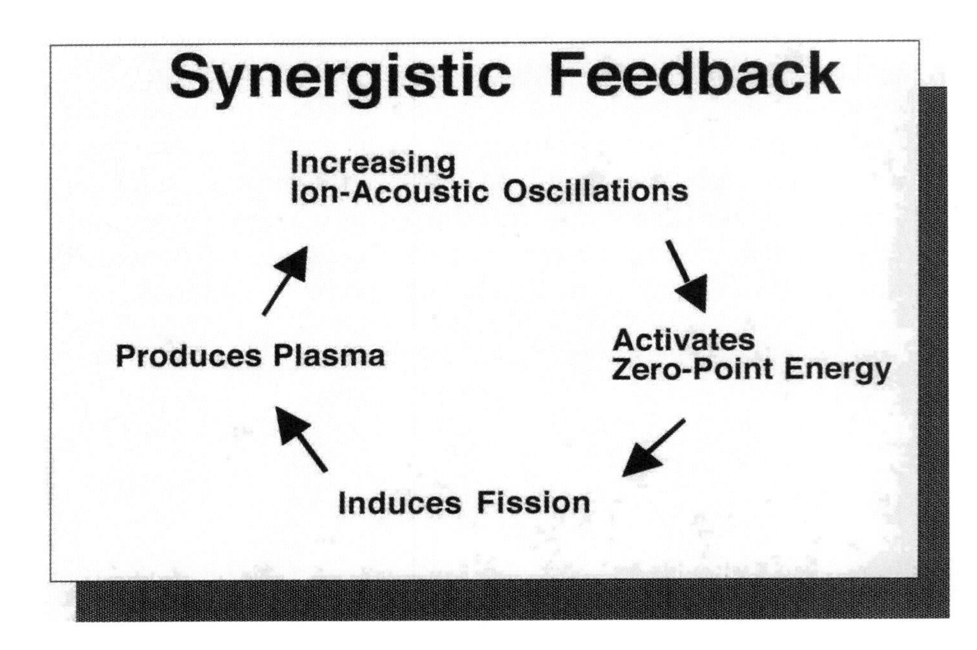
Synergistic Feedback
Increasing
Ion-Acoustic Oscillations
Activates
Zero-Point Energy
Induces Fission
Produces Plasma

20. Synergistic feedback. An example of interactive energy feedback is illustrated: Ion-acoustic plasma oscillations may activate a zero-point energy coherence, which could induce more fission, which can produce more plasma, which repeats the process by increasing the ion activity. Coherent coupling with zero-point energy opens new possibilities for synergistic energy interactions within a system.

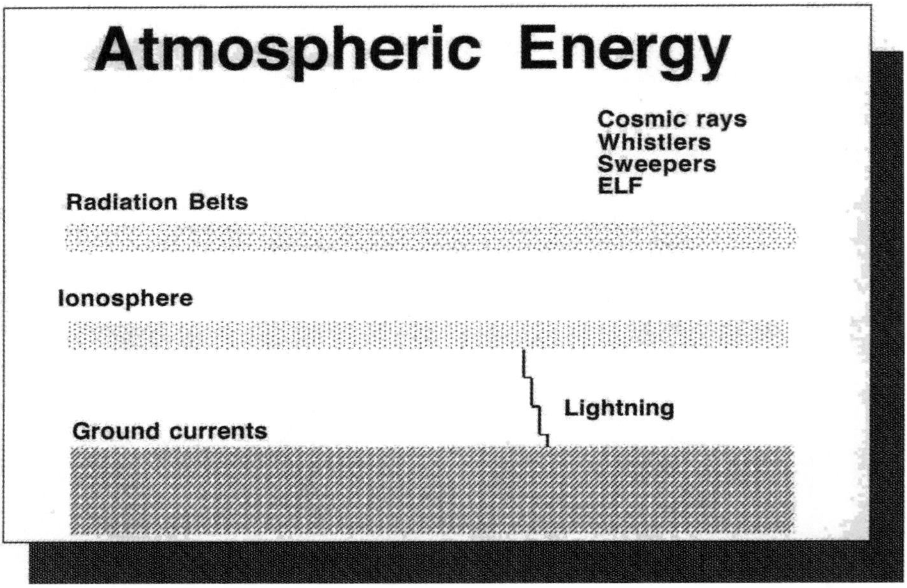

Atmospheric Energy

Cosmic rays
Whistlers
Sweepers
ELF

Radiation Belts

Ionosphere

Lightning

Ground currents

21. Atmospheric energy. There is certainly energetic activity occurring in the atmosphere, ionosphere, and radiation belts. It includes lightning, ground currents, whistler waves, sweeper waves, solar wind and cosmic rays. Whistlers and sweepers both rapidly change their frequencies and arise in the ionosphere plasma and radiation belts. A broadband detector (like Moray used) is needed to absorb such waveforms, but is there a sufficient concentration of energy to explain an output of kilowatts, especially in the final device where the antenna and ground were unnecessary?

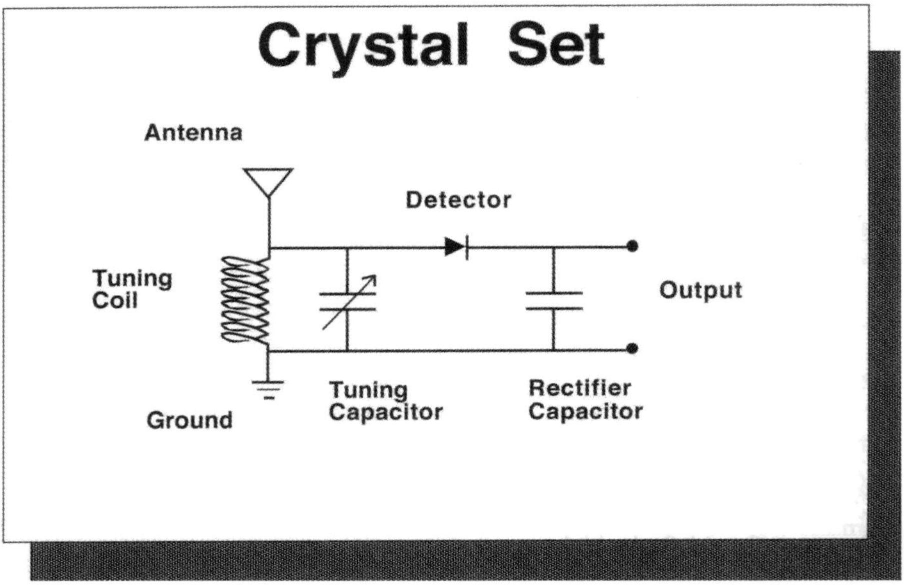

Crystal Set

Antenna

Detector

Tuning
Coil

Output

Tuning
Capacitor

Rectifier
Capacitor

Ground

22. Crystal set. Crystal sets were popular with hobbyist in the early years of electronics. The antenna and ground connect to a variable capacitor and inductor for tuning in the desired radio station. The rectifying crystal, typically a point-contact diode was used to detect the amplitude of the radio signal. Moray discovered that by removing the front end tuning capacitor, he effectively created a broadband low pass filter, which would pick up surges of energy from the environment. Moray focused much effort on improving the crystal detector to better rectify (gate) and amplify the incoming waves.

Moray's Detector

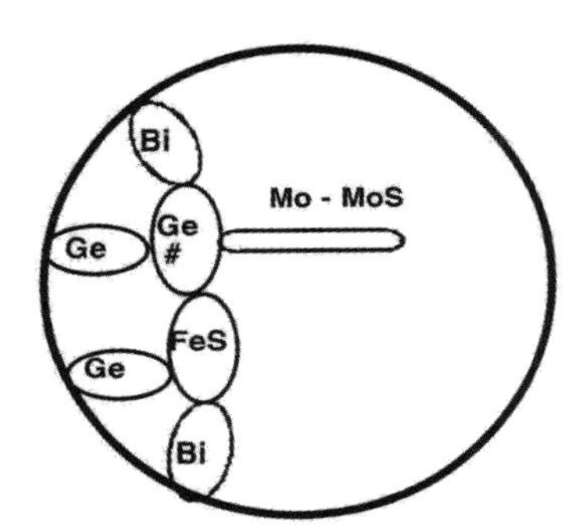

Bi - Bismuth
Mo - Molybdenum
MoS - Molybdenum Sulfide
FeS - Iron Sulfide
Ge - Germanium
\# + Zinc Sulfide
Radium
Uranium
Thorium

The Sea of Energy, Cosray Research Inst, Salt Lake City, 1978, p70.
1960, p130.

23. Moray's detector. An example of Moray's detector technology appeared in his patent application (Moray, 1960). Moray worked with the appropriate transistor materials (germanium, bismuth) in the 1920's well before Bell Labs discovery of the transistor in the late 1940's. Moray strived to maximize the glow plasma on the surface of the pellets by use of metallic (iron, molybdenum, zinc) sulfides, as well as using radioactive materials. Moray studied the text of Rutherford who noted that a mixture of radium, uranium and thorium provided more radioactivity than any one alone (Sego, 1981). The radioactive emissions induce luminescence in the metallic sulfides, which help maintain the surface plasma. It is the plasma, which provided amplification to the incoming signal so much so that Moray's crystal radio set could drive a loud speaker without any other power input.

Radioactive Cold Cathode

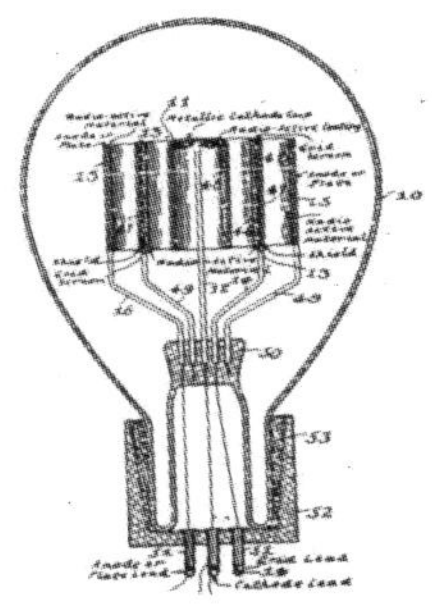

McElrath, U.S. Patent 2,032,545

1936 (Filed 1931)

24. Radioactive cathodes. In the early years of electronics it was popular to experiment with radioactive materials, and they were readily available through chemical suppliers. Often such material was used in cathodes to augment electron emission. Numerous patents (McElrath, 1936) were issued claiming this point, yet the patent office rejected Moray's energy invention on the grounds that he did not heat the cathodes in his tubes.

Cold Cathode Discharge Tube
Cupped Electrodes

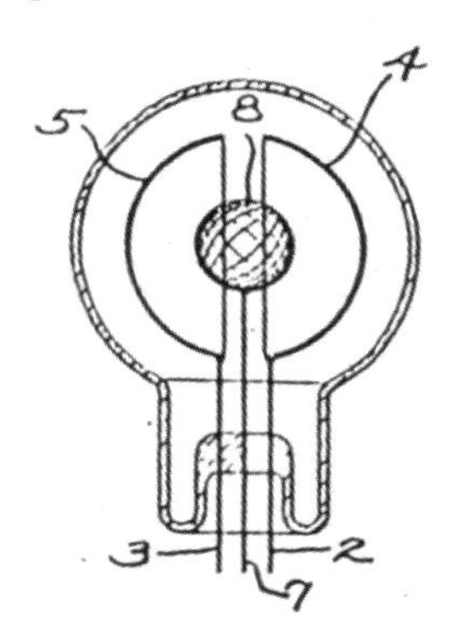

Farnsworth, U.S. Patent 2,184,910 (1939)

25. Fusor patent. Philo Farnsworth (1939), the inventor of television, was issued a patent that combined radioactive material with specially cupped electrodes to concentrate plasma. The device produced such unusually large power that Farnsworth believed it came from fusion. Could his invention be another example of plasma induced, zero-point energy coupling?

Resonant Nuclear Battery

Brown, U.S. Patent # 4,835,433

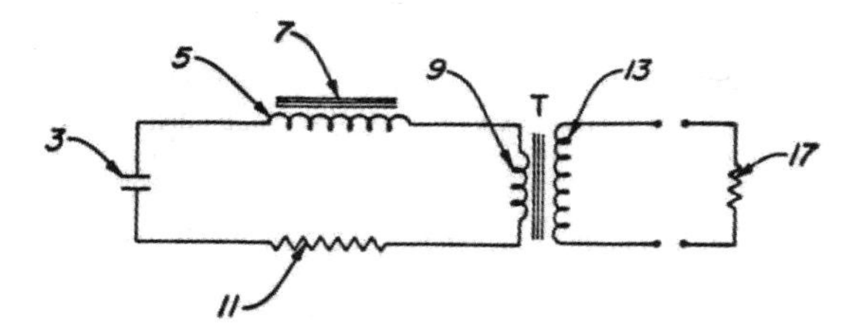

Radioactive core (7) produces cold plasma.

Circuit resonance must match ion-acoustic resonance.

26. Brown's resonant nuclear battery. The late Paul Brown's research perhaps contributes the most insight into understanding Moray's operating principle. Brown (1997) showed that a simple resonant (capacitor-inductor) tank circuit could be made self-running by bombarding the inductor coil with radioactive emissions. Excess energy would be produced, which Brown rectified to create the D.C. output for his battery. Does the nuclear radioactivity provide all the energy, or does it catalyze plasma whose ionic oscillations activate the zero-point energy?

Glow Plasma Oscillator

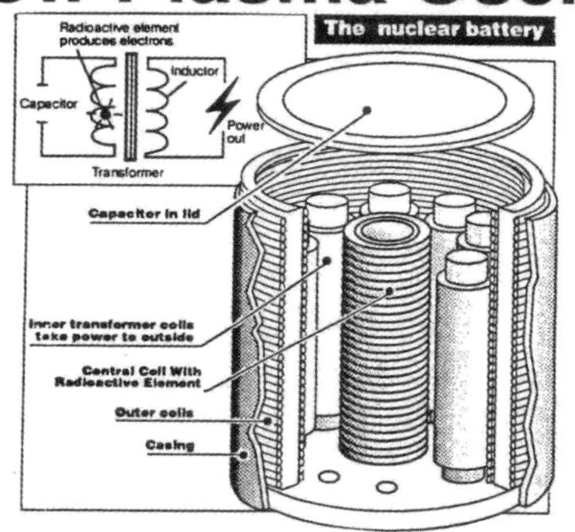

Paul Brown, U.S. Patent 4,835,433 (1989)

27. Hubbard device. Paul Brown made history in the new energy field in two ways: 1) He was the first to successfully replicate a self-running energy invention from a historic inventor, Alfred Hubbard (1919), and 2) he came closest to anyone in the field to breaching the marketplace. Brown's patent is essentially a detailed description of the Hubbard device, where the inductor coils are arranged in a circular configuration that some researchers hypothesize might induce vortex action in the ether. Brown wisely claimed his device was strictly an efficient nuclear battery that could produce five watts from a weak (one curie) radioactive source such as krypton 85 or strontium 90. In 1990 I privately conferred with Paul Brown and asked him how many people question that claim, since even assuming 100% conversion of all mass to energy, such weak radioactivity could provide only five milliwatts at best. The output is 1000 times too much. He answered about one in a hundred people would recognize the issue, but there are two even more important points. Brown was actually trying to create a 100-watt unit from the same technology. It was unstable, and sometimes there occurred surges of power, perhaps on the order of kilowatts, that would burn up the wires. He could stabilize the unit at five watts, and his company decided to sell that in order to capitalize further research. General Electric, in their due diligence for partnership to manufacture the battery, sent their nuclear physicist to investigate it. Paul Brown said the physicist lost sleep for a week because he could not explain the excessive energy. It was after this investigation that the suppression problems really began for Paul Brown, for it appears he successfully created a surprisingly simple, self-running, zero-point energy device.

Zero-Point Energy: Basis

Quantum Effects - Boyer, Phys. Rev. D 11(4), 2832 (1975).

Hydrogen Atom - Puthoff, Phys. Rev. D 35(10), 3266 (1987).

Energy Source - Cole, Puthoff, Phys. Rev. E 48(2), 1562 (1993).

Gravity - Puthoff, Phys. Rev. A 39(5), 2333 (1989).

Inertia - Haisch, Puthoff, Rueda, Phys. Rev. A 49(2), 678 (1994).

28. **Zero-point energy, the basis.** Although not widely studied in the engineering community, the zero-point energy is a profound and considerably complex research topic for physicists. The energy manifests as chaotic, highly energetic, electric field fluctuations inherent to the fabric of space. The name "zero-point" refers to absolute zero degrees Kelvin meaning it is not heat radiation, and it is what comprises pure empty space when all matter and radiation (heat, light, etc.) are absent. The fluctuations are so energetic that some physicists theorize it is the foundational basis for all particles and fields, sustaining their existence (Senitzky, 1973). The physics literature supports the idea that the ZPE is the basis for quantum effects (Boyer, 1975), atomic stability (Puthoff, 1987), gravity (Puthoff, 1989), and inertia (Haisch, 1994). In short it is the modern term for the ether, but unlike the static model of the 18th century ether, the zero-point energy offers opportunities for dynamic interaction.

Uncertainty Principle

$\Delta E \Delta t > \hbar$

Pair Production

29. Uncertainty principle and pair production. The ZPE was discovered theoretically as a term in the equations of quantum mechanics. It provided the underlying jitter for the Heisenberg uncertainty principle, and Dirac (1930) interpreted it as inherent to space where virtual (short lived) electron-positron pairs would pop in and out of existence in a chaotic maelstrom of energetic turbulence.

Single Vacuum Fluctuation

Time 10^{-43} sec

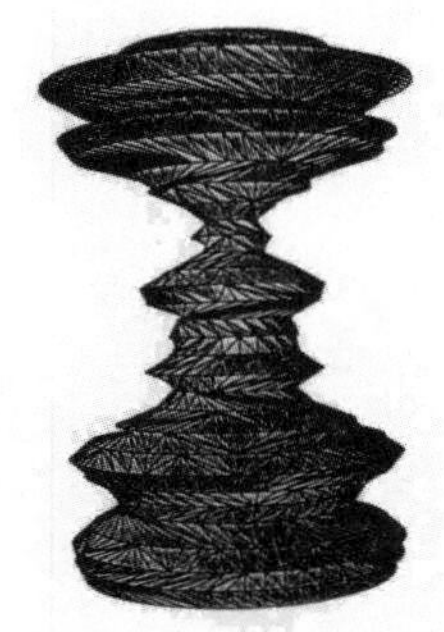

Space 10^{-33} cm

Lee Smolin, <u>Three Roads to Quantum Gravity</u>, 2001

30. Quantum gravity vacuum fluctuation model. Physicists are attempting to combine the theories of general relativity and quantum field theory to yield the ultimate unified theory known as quantum gravity. Recent successes in this area (which include and augment string theory) have attracted many physicists into participating. The foundation of quantum gravity is the energetic vacuum fluctuations, which organize to manifest both space-time and matter. Detailed modeling has been done at the Planck length, 10^{-33} cm. Shown here is a computer graphic generated from modeling a single vacuum fluctuation. It is from Lee Smolin's popular book (2001), *Three Roads to Quantum Gravity.*

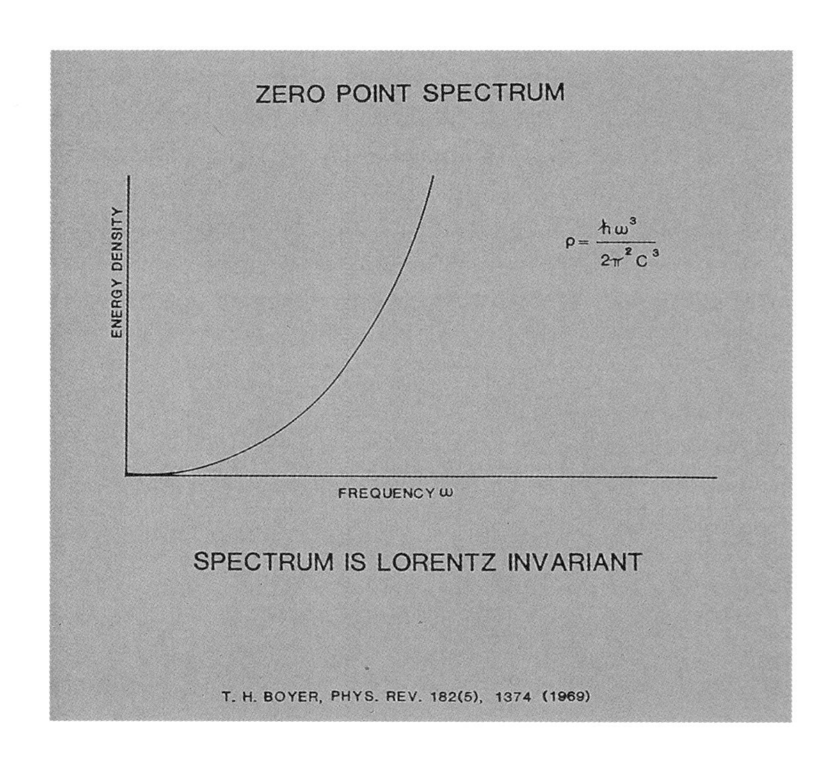
ZERO POINT SPECTRUM
ENERGY DENSITY
$\rho = \frac{\hbar \omega^3}{2\pi^2 C^3}$
FREQUENCY ω
SPECTRUM IS LORENTZ INVARIANT
T. H. BOYER, PHYS. REV. 182(5), 1374 (1969)

31. ZPE Spectrum. The spectrum of the vacuum fluctuations shows how the zero-point energy density varies with frequency. Theorists have derived the spectrum by applying the principle of Lorentz invariance from special relativity. All inertial frames (observers moving at constant velocity in free space) must observe the same ZPE spectrum. There is only one functional form that fulfills Lorentz invariance: the energy density must be proportional to the cube of the frequency. Boyer (1975) showed that many results in quantum mechanics (e.g. a blackbody radiator) historically derived by assuming discrete energy states, can instead be derived from this particular ZPE spectrum. The successes from using this theoretical approach established the field of stochastic electrodynamics, which strives to explain quantum effects by modeling matter's interaction with the vacuum's zero-point fluctuations. However, a philosophical problem immediately arises from the ZPE spectrum: The energy density increases without bound as the frequency increases. Without introducing a frequency cutoff to the model, the energy density in space appears infinite.

32. Warped Space. Wheeler (1962) theory of geometrodynamics derived a "natural" cutoff for the zero-point energy spectrum using the theory of general relativity. As mass or energy density increases, space-time curvature occurs. The figure illustrates how the path of light bends in response to a gravitating body warping the fabric of space.

Wormholes

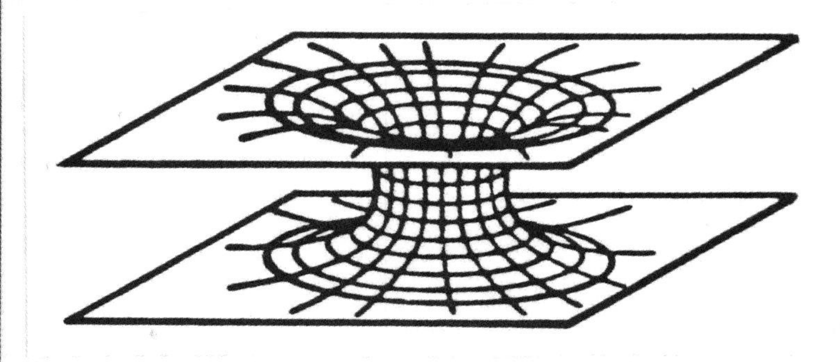

Wheeler, Geometrodynamics, Academic Press, NY, 1962

Hawking, Phys. Rev. D 37(4), 904 (1988)

33. **Wormhole.** With sufficient energy density space can pinch off to form a "wormhole," which can interconnect one region of 3-space with another. The regions can be in parallel universes or possibly remote locations of the same universe. Wheeler describes a "superspace" containing an infinite number of three-dimensional universes. Microscopic ZPE wormholes of electric flux interconnect the universes. The wormholes are sized at the Planck length, 10^{-33} cm, and contain an extraordinary (mass equivalent) energy density: 10^{94} g/cm^3.

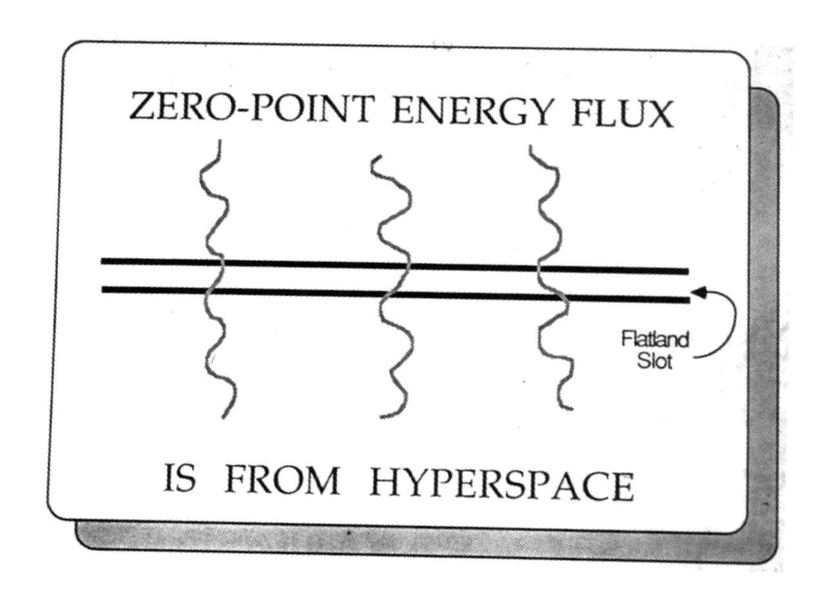
ZERO-POINT ENERGY FLUX
Flatland
Slot
IS FROM HYPERSPACE

34. Flux from Hyperspace. Wheeler's geometrodynamics models the zero-point energy as an orthogonal electric flux from hyperspace intersecting our 3-space. In the figure the thin "flatland" slot represents our three dimensional space. The slot thickness is related to Planck's constant. An enormous ZPE flux passes directly through at right angles, yet it is barely detectable because so little is aligned parallel to our 3-space.

THE ZERO-POINT ENERGY MAY ARISE FROM AN ORTHOGONAL ELECTRIC FLUX FROM THE FOURTH DIMENSION

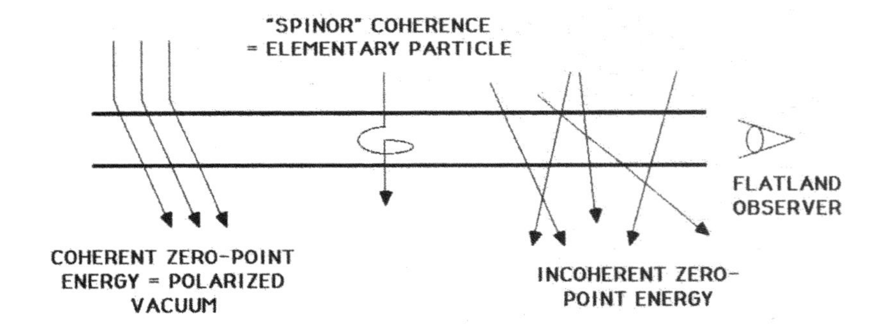

"FLATLAND SLOT" REPRESENTS THREE-DIMENSIONAL SPACE, SLOT WIDTH IS RELATED TO PLANK'S CONSTANT

35. Orthogonal Flux Model. Wheeler (1962) suggests that the orthogonal ZPE flux is the foundation for all matter and energy in our universe. As the flux passes through any jitter aligned in 3-space is detected as the background zero-point fluctuations. If there is a slight tilt to the penetrating flux, a net vector component aligns in our space, and it would manifest as vacuum polarization. If there is a vortex in the flux, we detect is as an elementary particle. An analogy is that the elementary particle is like a whirlpool whose existence is sustained by the flow of the (ZPE flux) stream. Wheeler's orthogonal flux model is perhaps the most powerful of all the zero-point energy descriptions in today's physics literature.

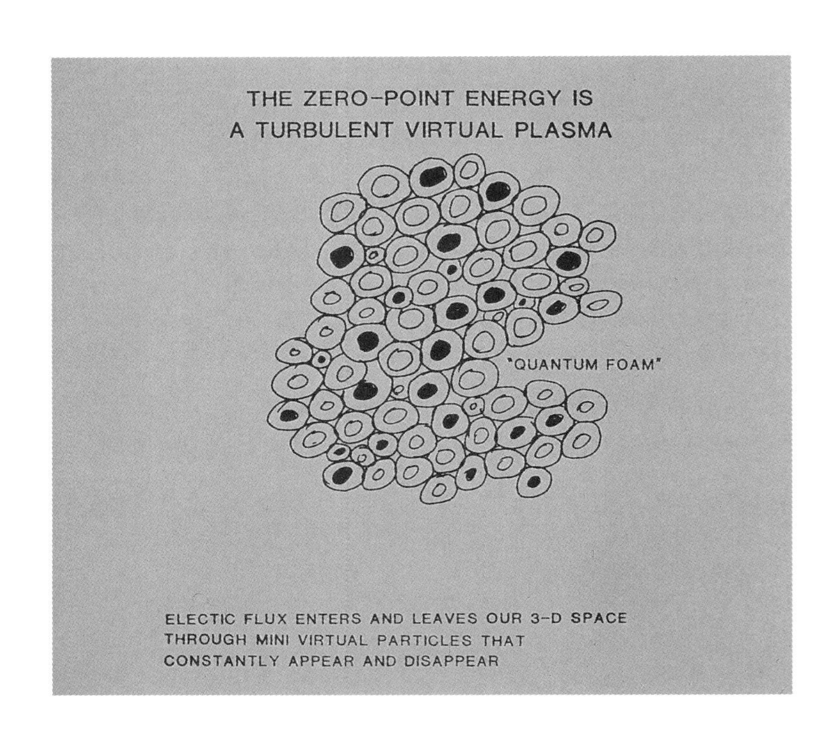
THE ZERO-POINT ENERGY IS
A TURBULENT VIRTUAL PLASMA
"QUANTUM FOAM"
ELECTIC FLUX ENTERS AND LEAVES OUR 3-D SPACE
THROUGH MINI VIRTUAL PARTICLES THAT
CONSTANTLY APPEAR AND DISAPPEAR

36. Quantum foam. Wheeler's geometrodynamics describes how the penetrating ZPE electric flux manifests in our three dimensional space. The flux enters through microscopic channels (wormholes) and yields a turbulence of extraordinarily small (10^{-33} cm) mini white holes (flux entering) and mini black holes (flux leaving). The mini holes are like (sub quantum) positive and negative charges. Wheeler calls the resulting chaotic turbulence of the fabric of space the "quantum foam." The energy density of flux passing through the quantum foam is enormous (a mass equivalence of 10^{94} g/cm^3). The quantum foam model for the ZPE is somewhat similar to turbulent plasma. Can net energy be extracted from such activity?

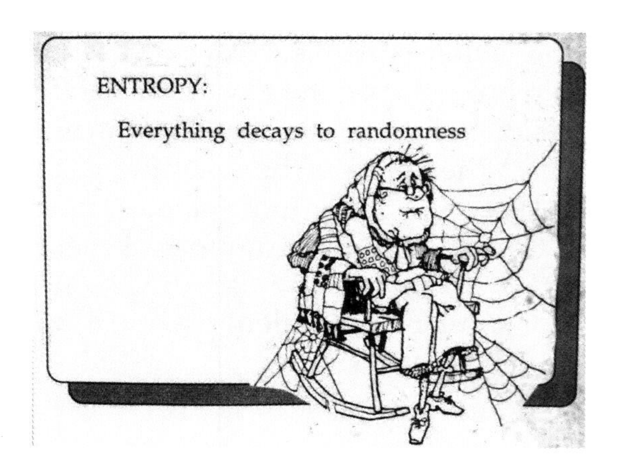
ENTROPY:
Everything decays to randomness

37. Entropy. Can coherent order arise from chaos? At first the answer appears to be no. The common understanding of the law of entropy is that chaotic behavior must always remain random and would never self-organize.

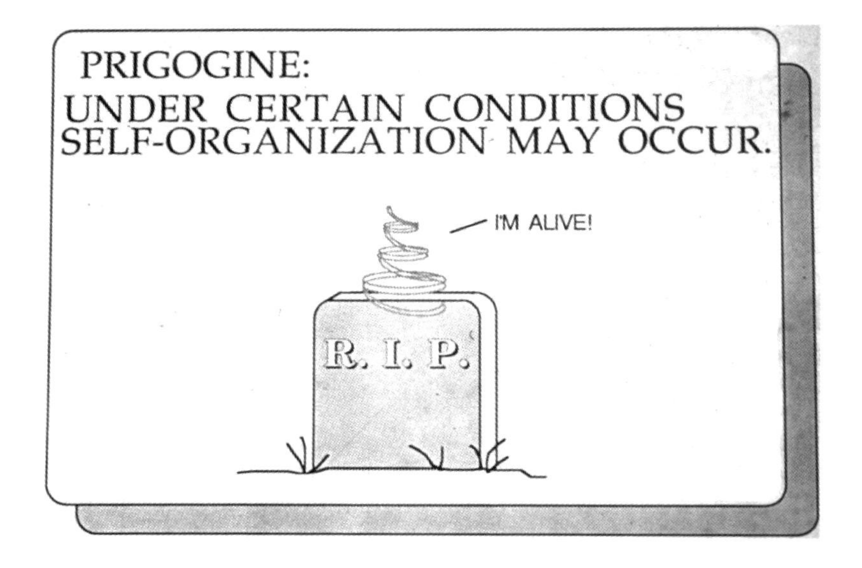
PRIGOGINE:
UNDER CERTAIN CONDITIONS SELF-ORGANIZATION MAY OCCUR.
I'M ALIVE!
R. I. P.

38. Prigogine. However, in 1977 Ilya Prigogine won the Nobel Prize in chemistry for identifying under what circumstances a system could evolve from chaos toward self-organization. Prigogine (1977, 1984) used general system theory, and showed that any chaotic system that exhibited the appropriate characteristics could potentially self-organize. (Suzuki, 1984, et al.)

System Self-Organization

- Nonlinear
- Far From Equilibrium
- Energy Flux

39. Principles for self-organization. A system must exhibit three requirements in order for it to self-organize: 1) It must be nonlinear, 2) far from equilibrium, and 3) have an energy flux passing through it. The theoretical models describing the zero-point energy fulfill these requirements. Merging the theories of the zero-point energy with the theories of system self-organization open the scientific possibility of activating a coherent ZPE interaction, which could become a basis for new technology (King, 1989, 2001).

Principles For Cohering The Zero-Point Energy

- Highly Nonlinear System
- Abruptly Driven Far From Equilibrium
- Maximize ZPE Interaction Using

    > Ions

    > Bucking Fields

40. Principles for ZPE coherence. Prigogine's requirements point the way for inventing a system that could tap the zero-point energy: 1) Work with a highly nonlinear system like a plasma, 2) drive it far from equilibrium by an abrupt discharge, and 3) work with the appropriate elementary particles which maximize their influence when interacting with the zero-point energy. There are examples in the literature of each of these principles in action.

Nonlinear Electric Dipole Oscillator Absorbs Zero-Point Energy

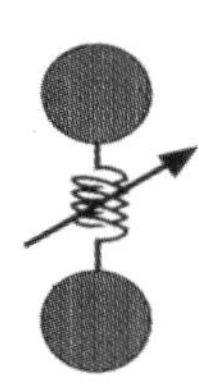

Boyer, Phys. Rev. D 13 (10), 2832 (1976)

41. Nonlinear dipole. In 1976 Timothy Boyer, the leading ZPE physicist in the United States at that time, published an analysis of a nonlinear electric dipole interacting with the ZPE (Boyer, 1976). He was surprised that the equations predicted particular modes of the dipole's oscillations would amplify and absorb energy straight from the vacuum fluctuations. Since he was unaware of Prigogine's research, he criticized the result in his conclusions for he did not believe it was possible to tap random fluctuations. Had he believed his own theoretical derivation, he could have been the first to predict the possibility of vacuum energy extraction via nonlinear interaction.

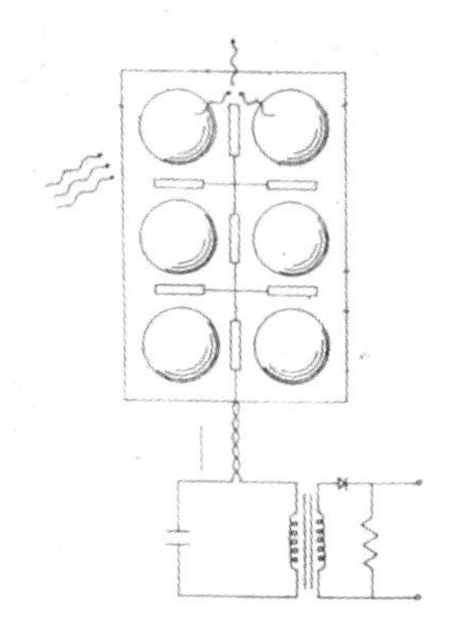

Nonlinear Dielectric Spheres

Resonate at High Frequency

Each Pair Slightly Detuned

Emits Low, Beat Frequency

Circuit Absorbs Beat Frequency

Mead, U.S. Patent # 5,590,031 (1996)

42. Microscopic nonlinear dipoles. Instead, Frank Mead (1996) was awarded a patent for extracting vacuum energy via nonlinear dipoles. Small nonlinear dipoles embedded in a substrate similar to computer chips resonate with the high frequency, highly energetic modes of the ZPE. Useful energy is extracted at the lower beat frequency via standard electronic means. Mead's patent is an example of using straightforward, solid state engineering to tap the vacuum energy.

Abrupt Motion of Matter Activates Vacuum Energy

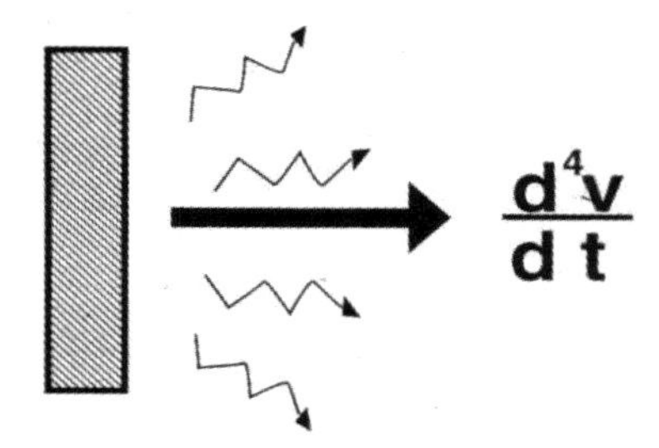

Barton, Eberlein, Ann. Phys. 227, 222 (1993)

43. Abrupt motion. Claudia Eberlein (1996) from Cambridge University published her thesis on the interaction of macroscopic matter with the zero-point energy. She showed that the abrupt motion of a wall of matter coherently activates photons directly from the vacuum. The more abrupt the motion the better, for the activation is proportional to the fourth derivative of velocity.

Sonoluminescense

Photon emissions faster than atomic transitions.
Energy amplified by 100 billion.

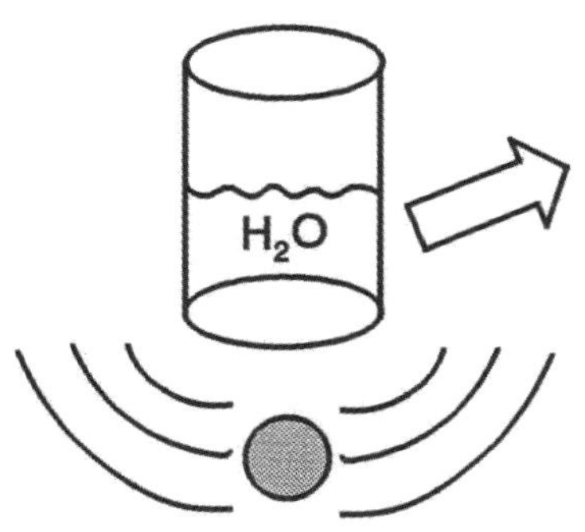

Ultrasonic Excitation

Barber, Putterman, Nature 353, 318 (1991)

Sonoluminesence

Activates Vacuum Energy

Appears 40,000 Degrees K

High Frequency Only

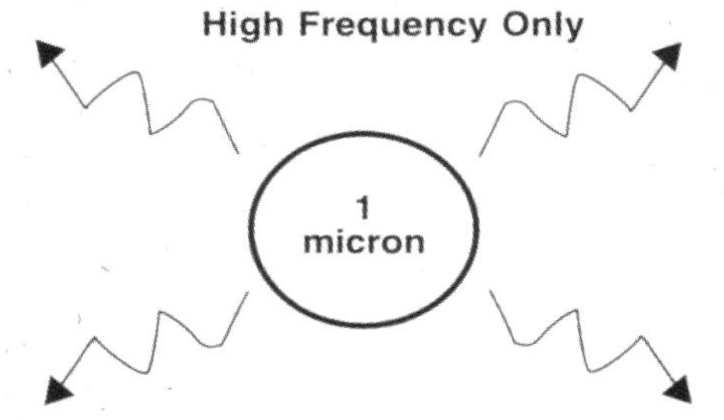

Bubble Boundary Resonance Yields 1000 X Gain

Eberlein, Phys. Rev. Lett. 76, 3842 (1996)

44. Sonoluminesence. Eberlein applied her result to explain sonoluminesence, where water mixed with inert gas under ultrasonic stimulation emits a bluish light, which cannot come from atomic transitions (Barber, 1991), but instead results from the abrupt compression during bubble collapse converting vacuum energy to light. The abrupt motion of matter is a vacuum energy activator.

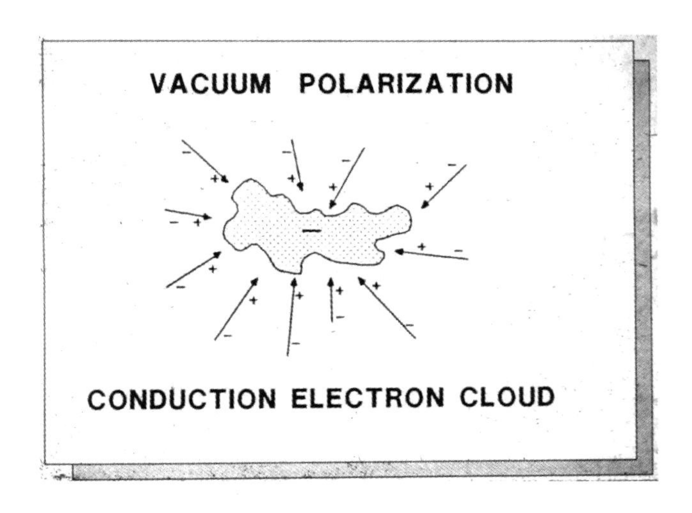
VACUUM POLARIZATION
CONDUCTION ELECTRON CLOUD

45. Vacuum polarization of electron. How do elementary particles activate zero-point energy? Quantum electrodynamics describes the interaction of elementary particles with the ZPE as "vacuum polarization." The different elementary particles have different vacuum polarization characteristics (Scheck, 1983). Electrons, especially those within a metal's conduction band, exhibit a smeared cloud-like characteristic that is essentially in equilibrium with the vacuum fluctuations (Senitzky, 1973). Standard electrical wires and antennas would make poor transducers to detect or activate a net energy from the ZPE.

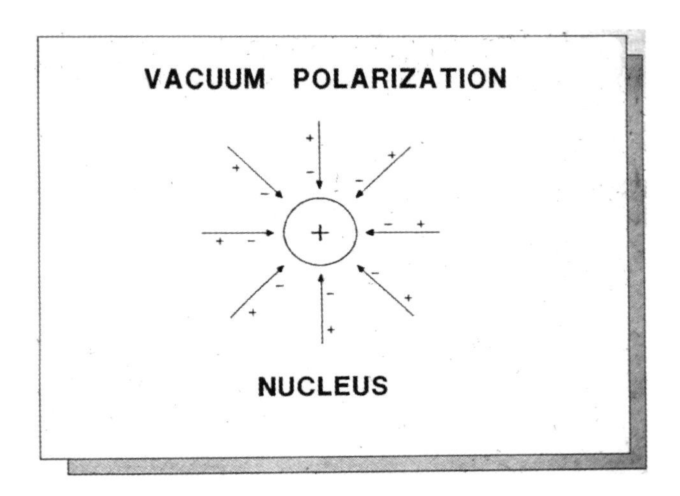
VACUUM POLARIZATION
NUCLEUS

46. Vacuum polarization of nucleus. The vacuum polarization surrounding atomic nuclei exhibit steep convergence of field lines manifesting an organizing influence on the vacuum fluctuations. This suggests that the key for activating ZPE coherence is abrupt motion of atomic nuclei.

Exotic, Coherent Vacuum States
in Quantum Electrodynamics
arise from heavy ion collisions.

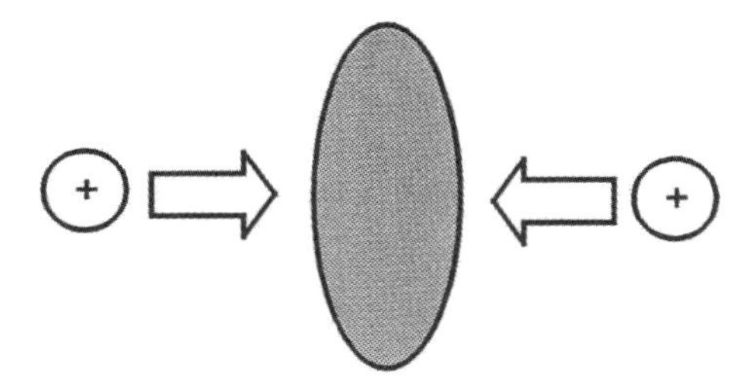

Celenza, et al., Phys. Rev. Lett. 57(1), 55 (1986).

47. Exotic coherent vacuum states. Particle accelerator experiments that collide atomic nuclei do indeed create exotic, coherent vacuum energy states (Celenza, 1986, et al.). In these experiments there is no searching for a net energy gain since the majority of the physics community believe that all the energy comes from the accelerator. Here it would be difficult to measure a net energy since there are so many losses associated with collision events.

Macroscopic Vacuum Polarization
Displacement Currents

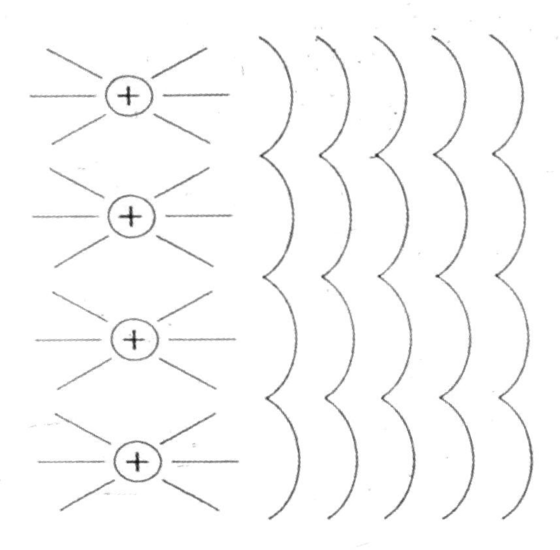

Ion acoustic oscillations of a plasma

Plasma Ion Acoustic Mode

- Large radiant energy absorption
- High frequency spikes
- Runaway electrons
- Anomalous plasma heating
- Anomalous plasma resistance

48. Ion-acoustic oscillations. If the abrupt motion of a single nucleus can activate the vacuum energy, what if we move a large number together? This is exactly what happens during ion-acoustic resonance of plasma. Numerous positive ions move synchronously, and plasma experiments have manifested energetic anomalies including high-frequency voltage spikes, anomalous heating (Kalinin, 1970, et al.), anomalous resistance, and "run-away" electrons (Sethian, 1978, et al.). Synchronous abrupt motion of plasma ions appears to be an engineering key to activate and couple vacuum energy to the plasma. T.H. Moray stressed that ion oscillation was the fundamental mode occurring in his tubes of his energy machine.

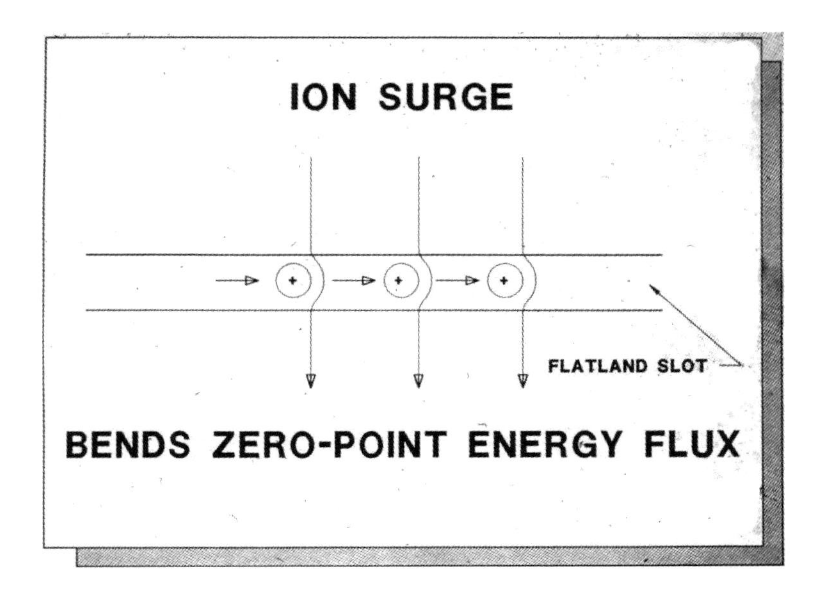

Feb. 1, 1949. T. H. MORAY 2,460,707

ELECTROTHERAPEUTIC APPARATUS

Filed April 30, 1943 3 Sheets-Sheet 1

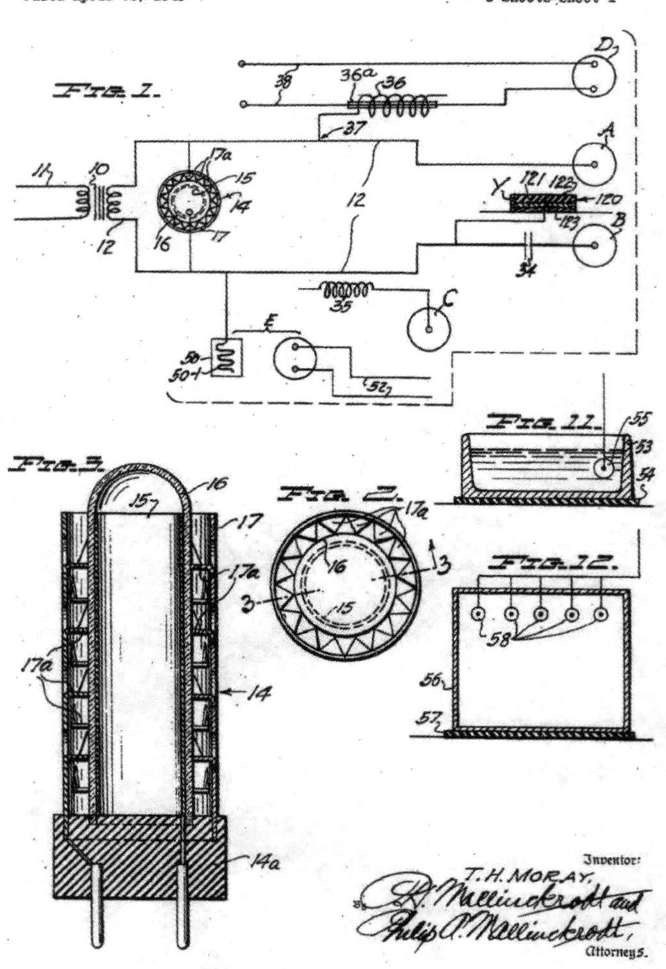

49. Therapeutic apparatus patent. In the 1930's the U.S. patent office rejected Moray's original application for his energy machine because the examiner could not understand how the device could output so much energy without heating the cathodes of his tubes. However, in 1949 they granted Moray patent 2,460,707 for a therapeutic apparatus. The patent contains plasma tubes and illustrates the craftsmanship and engineering skill that Moray employed to produce and control the corona within his tubes. A careful study of this patent yields a surprise: Three of the tubes (patent figures 14-19) do not "fit" the therapeutic device. Instead analysis of these tubes shows they functionally match the oscillator and valve tubes that would be part of the energy device. It appears that Moray (1949) was attempting to cover in this patent some critical components of his energy device.

Feb. 1, 1949. T. H. MORAY **2,460,707**

ELECTROTHERAPEUTIC APPARATUS

Filed April 30, 1943 3 Sheets-Sheet 2

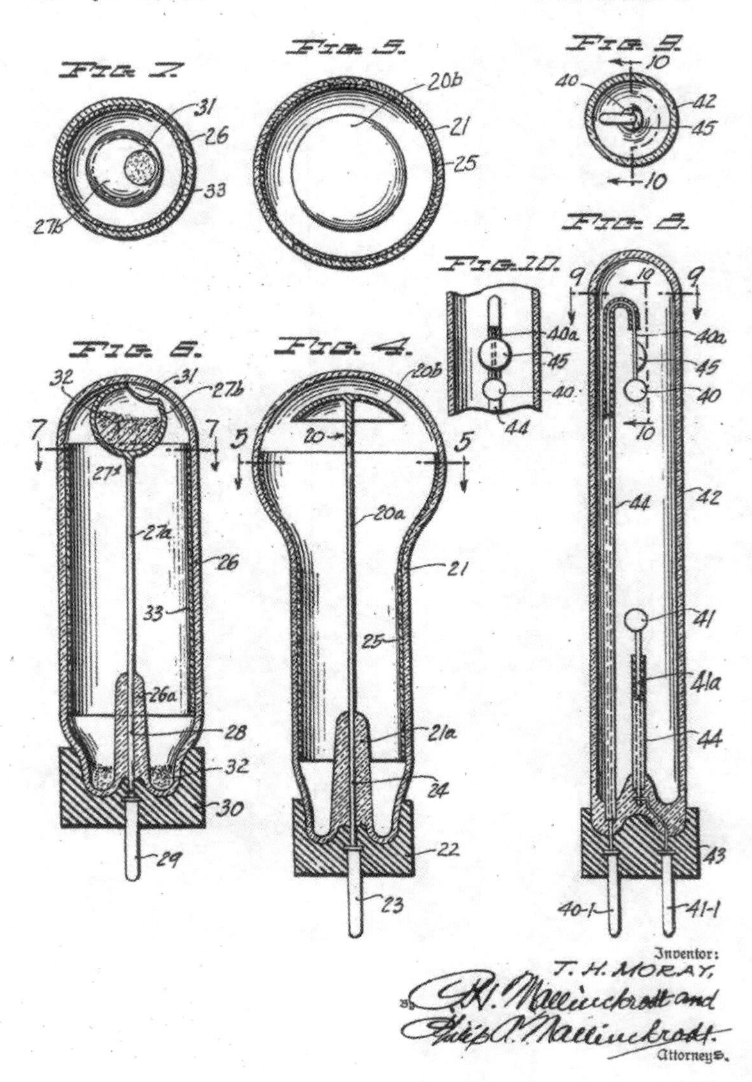

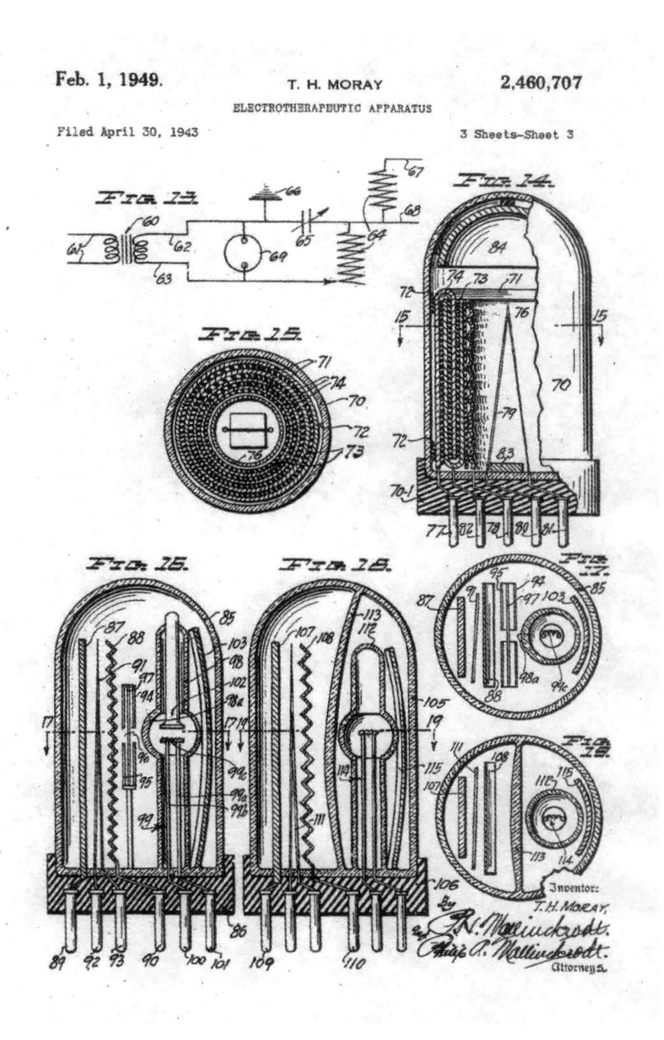
Feb. 1, 1949.
T. H. MORAY
2,460,707
ELECTROTHERAPEUTIC APPARATUS
Filed April 30, 1943
3 Sheets-Sheet 3
Inventor:
T. H. MORAY,
Attorneys.

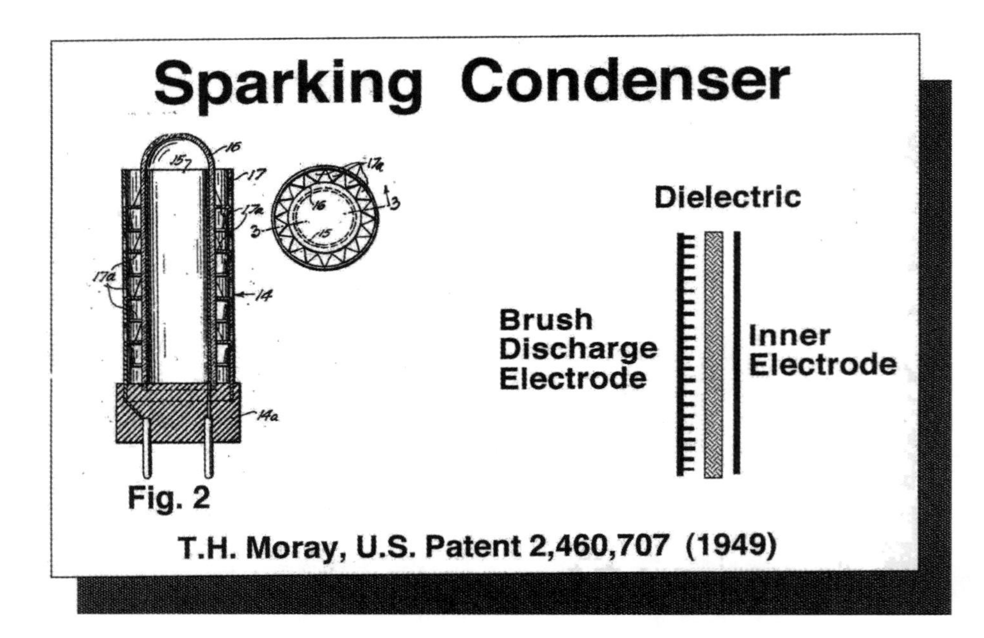
Sparking Condenser
Dielectric
Brush
Discharge
Electrode
Inner
Electrode
Fig. 2
T.H. Moray, U.S. Patent 2,460,707 (1949)

50. Sparking condenser tube. The first tube described (patent figures 2, 3) belongs to the therapeutic apparatus. It contains a cylindrical array of sharp pointed electrodes, which emit a brush discharge corona. Moray appears to have discovered that the ion oscillations in a corona might produce "emanations" which are therapeutically beneficial. One hypothesis is that the ion oscillations launch a macroscopic vacuum polarization wave (a subtle ZPE coherence) that interacts with the ions around a cell wall membrane, thus creating a biological influence at relatively low power. This could be effective if proper resonant frequencies were discovered that stimulated a cellular response appropriate for healing. The sparking condenser is an example of a relatively simple tube that manifests Moray's theme, "oscillate the ions."

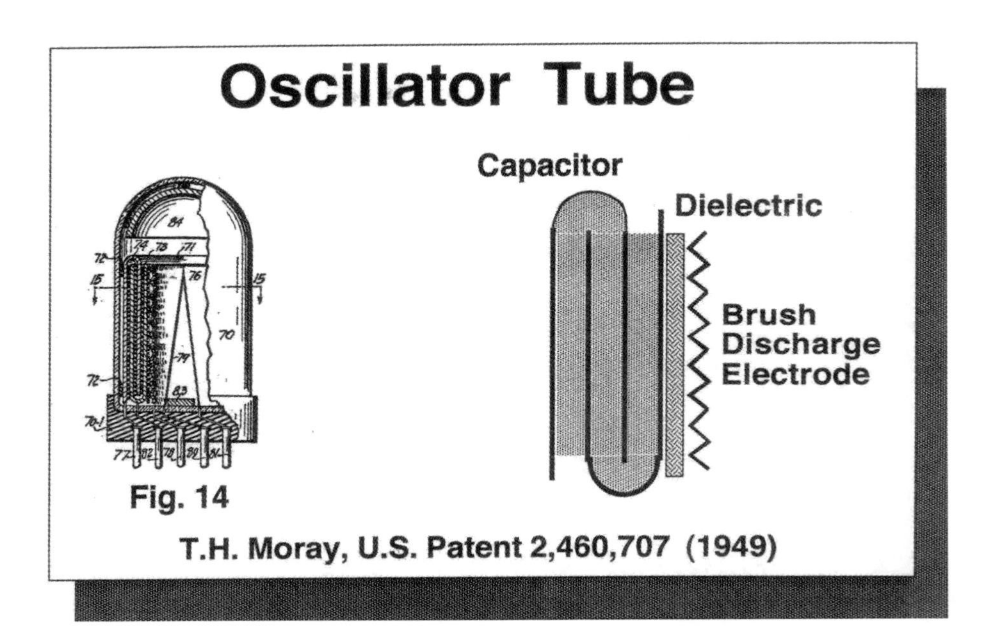
Oscillator Tube
Capacitor
Dielectric
Brush
Discharge
Electrode
Fig. 14
T.H. Moray, U.S. Patent 2,460,707 (1949)

51. Oscillator tube. The oscillator tube (patent figures 14, 15) appears to be the primary ion oscillator tube of Moray's energy device. He claimed such a tube exhibited extraordinary capacitance (one farad) while it was running at its particular resonant mode. The inner electrode is corrugated to support a brush discharge corona. A double wall, cylindrical capacitor surrounds the inner electrode. It contains a dielectric that presumably yields the large capacitance. In his book Moray mentioned the use of dielectrics such as powered quartz, and he showed a consistent pattern of mixing in radioactive materials such as radium salts and uranium ores. Typical dielectric materials do not have a sufficient dielectric constant to manifest the claimed capacitance. However, if one assumes that whenever Moray mentions a dielectric he could really be augmenting a dielectric powder with radioactive material, then interstitial glow plasma (microscopic corona between the grains of the powdered dielectric) could be activated at a low threshold voltage within the mixture. Since plasma exhibits an extreme electrical polarization, it can manifest a huge effective dielectric constant especially during its ion oscillations. Moray also mentioned that his tubes could contain inert gases, mercury vapor, moist vapor and radioactive material (which lowers the ionization voltage threshold). The tube supports ion-acoustic plasma activity from the corrugated electrode all the way through the dielectric of the cylindrical capacitor. Once the tube is undergoing ion oscillations, it not only could manifest a huge capacitance, it might also energetically couple directly with the zero-point energy.

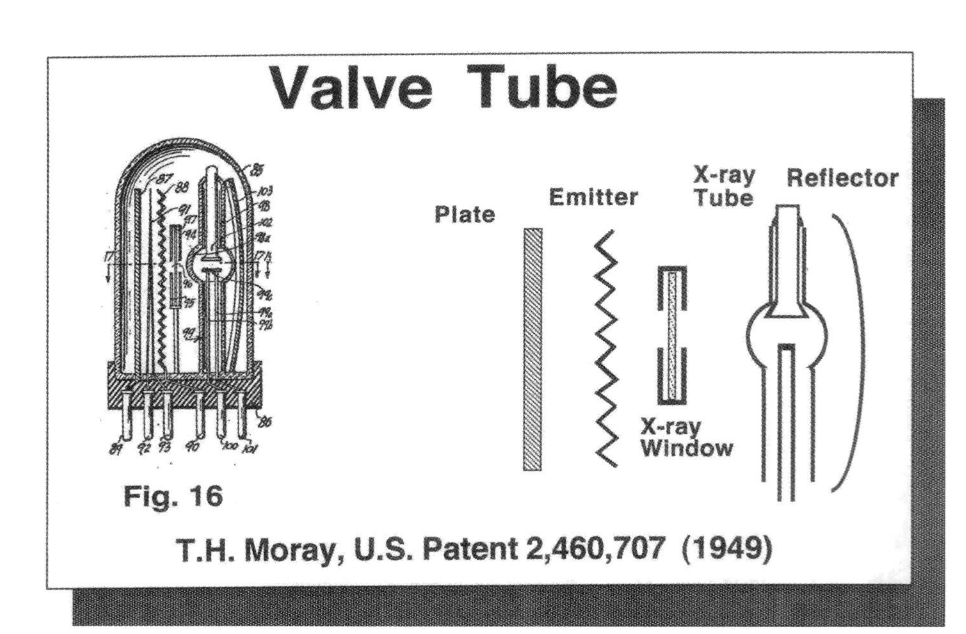
Valve Tube
Plate
Emitter
X-ray Tube
Reflector
X-ray Window
Fig. 16
T.H. Moray, U.S. Patent 2,460,707 (1949)

52. Valve tube. Moray's valve tube (patent figures 16, 17) is really a tube within a tube that acts like a triode switching tube. However, it is designed to switch a large unidirectional ion surge in response to a small discharge event triggered within the inner tube. The timing can be appropriately controlled by electronic means. The inner tube emits x-rays from electron collisions with its anode. The x-rays pass through the slot to trigger ionization in the channel between the anode and corrugated cathode of the outer tube to create the abrupt transient, ion surge. If the ion surge activates extra energy from the ZPE, its polarization pulse can provide amplification as it drives the oscillator tube. The valve tubes gate energy between the stages of the Moray circuitry, and each pulse from them offers a potential energy gain.

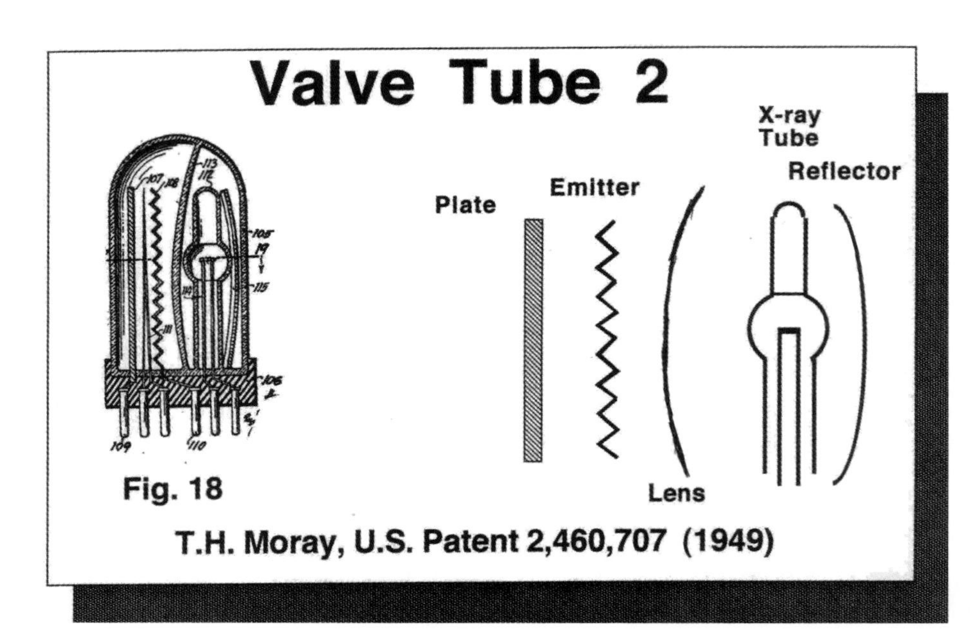
Valve Tube 2
X-ray
Tube
Reflector
Emitter
Plate
Lens
Fig. 18
T.H. Moray, U.S. Patent 2,460,707 (1949)

53. Valve tube 2. The second embodiment of the valve tube (patent figures 18, 19) operates similarly to the first. Here a lens and reflector are used to focus light and ultraviolet emissions from the inner tube to ionize the switching channel in the outer tube. In both valve tubes the corrugated electrode supports corona buildup just prior to the switching event. This seems to augment the energetic activity since more ions become available to participate in a synchronous surge.

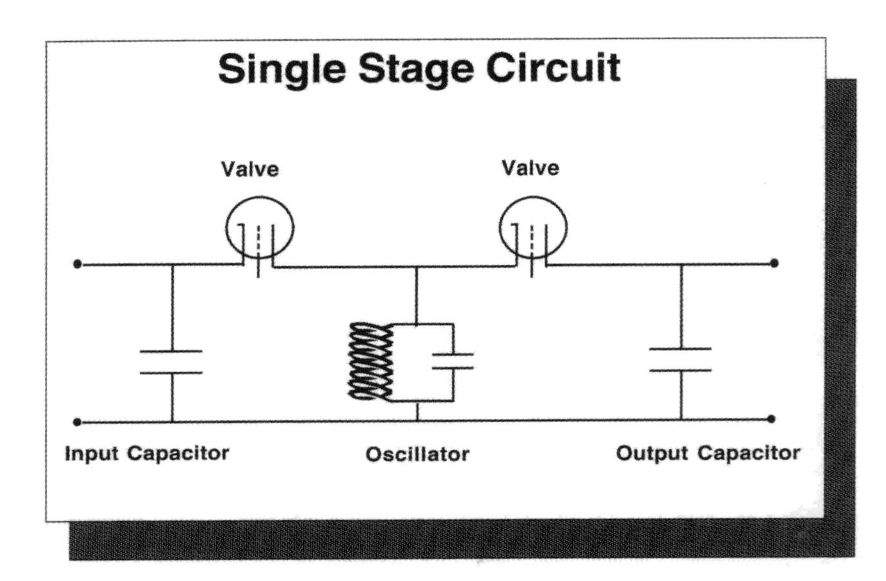
Single Stage Circuit
Valve
Valve
Input Capacitor
Oscillator
Output Capacitor

54. Single stage circuit. A single stage of the Moray's energy device illustrates the fundamental method for extracting energy. Energy stored on the input capacitor is switched through the input valve (tube) timed in phase to drive the oscillator tube. The oscillations gradually grow in voltage amplitude. When the voltage exceeds the appropriate threshold, the output valve (tube) is switched for an instant to pulse the output capacitor, which gradually charges. (The timing and control circuitry is not shown; it can be created by standard electrical engineering methods.) After many oscillating cycles the output capacitor reaches its full charge, at which point its energy must be dumped. Each pulse from the input valve tube drives the oscillator tube at its resonant frequency. Since radioactive material in the oscillator tube maintains its plasma, there are few losses and the amplitude readily grows. If synchronous ion motion activates ZPE, then the single stage circuit from the input valve tube to the output capacitor might manifest continuous, dynamic, vacuum polarization coherence in the spatial region surrounding the circuit components.

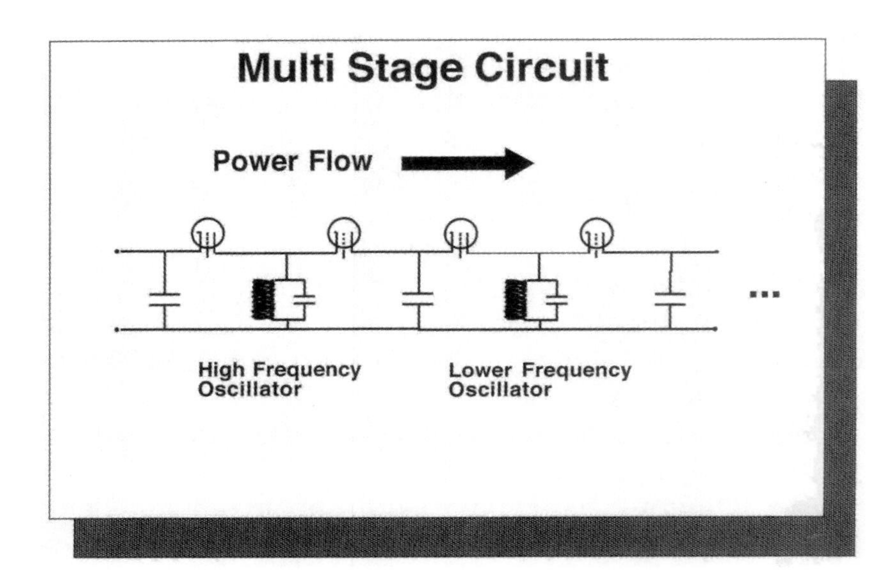
Multi Stage Circuit
Power Flow
...
High Frequency
Oscillator
Lower Frequency
Oscillator

55. Multistage circuit. Many single stage circuits can be cascaded to construct the multistage circuit. Each output capacitor becomes the input capacitor for the succeeding stage. A prior stage must oscillate at a considerably higher frequency than its subsequent stage. At each stage many input pulses must integrate on the output capacitor to yield one pulse that drives the next stage. If vacuum polarization pulses were guided along the conductors, the entire multistage circuit might manifest a large scale, distributed vacuum polarization coherence. If this is the primary form of energy channeled by Moray's device, it could then explain the cold current effect as well as the glass penetration experiment. Vacuum polarization waves that surround the thin wires are the primary form of energy transport, not electron conduction. Therefore ohmic heating is minimal. Moray appears to be the first inventor to engineer a novel form of "cold electricity."

Infinite Energy

Cold Fusion and New Energy Technology

Published by Cold Fusion Technology Vol.2, No.7, 1996

New Energy Electric Power Generation— Now!

© Correa & Correa, 1996

Dr. Paulo Correa and Alexandra Correa of Labofex in Canada with their PAGD Reactors (Pulsed Abnormal Glow Discharge). *First* to demonstrate Commercial-Grade Electricity Generation— a "New Energy" technology with *no thermal conversion required!*

PAGD REACTOR DRIVING ELECTRIC MOTOR

© Correa & Correa, 1996

Also in this issue...

- Japan research yields cold fusion transmutations in metal.
- Los Alamos Lab's cold tritium generator goes public.
- NASA confirms excess energy.

U.S. $5.95

Canada $7.95

07

7 50644 87813 7

56. Correa's pulse abnormal glow discharge (PAGD) tube. Paulo and Alexandra Correa (1995) appear to have invented a tube that works on the same principle as Moray's plasma tubes. In their PAGD tube, glow plasma gradually builds up on the cathode from a charging circuit. When the voltage reaches the breakdown threshold the tube discharges for an instant, and then the arc is quenched. Correa stressed that only the leading edge of the discharge event provides an energy gain. The arc that follows is standard electron flow, which creates heating losses. The charging circuit must be designed to stop the current flow immediately after the tube fires. Then the circuit gradually recharges the cathode glow plasma for the next cycle. The circuit parameters can be adjusted to pulse the tube at a controllable repetition rate (0.1 Hz - 1 KHz). Cathode size and geometry are important; the more glow plasma that accumulates prior to the discharge, the bigger the energy gain.

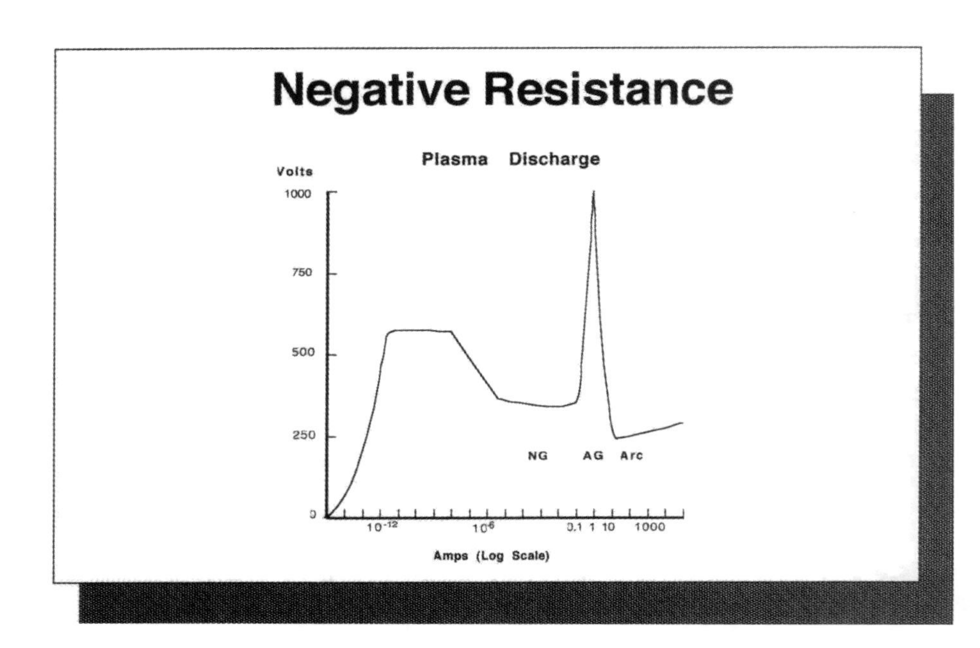
Negative Resistance
Plasma Discharge
Volts
1000
750
500
250
0
NG
AG
Arc
10^{-12}
10^{-6}
0.1 1 10 1000
Amps (Log Scale)

57. Negative resistance. Correa voltage-current plot of the discharge event identifies a negative resistance region (the downward sloping edge of the voltage spike). Negative resistance exhibits an energy gain. The circuit parameters should be tuned to operate the tube around this region. Correa has claimed to measure a four to one net energy gain. Also, he has made a dual circuit system involving two tubes and two batteries, where each PAGD circuit charges the other circuit's battery, and has run this for hours claiming no loss of battery power. He has not yet run the ideal test: Use capacitors instead of batteries to demonstrate a self-running system. Nonetheless Correa appears to have rediscovered the fundamental operating principle behind Moray's tubes with a relatively simple configuration that does not involve radioactive materials.

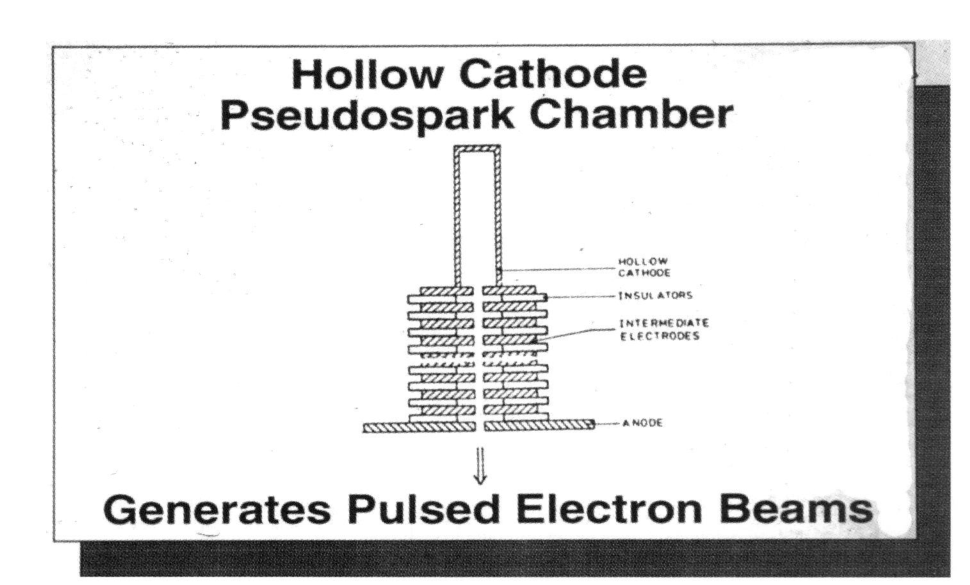
Hollow Cathode
Pseudospark Chamber
HOLLOW
CATHODE
INSULATORS
INTERMEDIATE
ELECTRODES
ANODE
Generates Pulsed Electron Beams

58. Pseudosparks. There is field of electrical engineering that uses technology similar to Correa's tube. Power engineers utilize high capacity switches involving hollow cathodes in which glow plasma accumulates prior to being switched (Gundersen, 1990). Typically, a laser triggers the switching event, and the hollow cathode then launches a "pseudospark" followed by the current. The advantage of using the hollow cathode is that large currents can be switched cleanly with a sharp rise time. There is an engineering conference every year dedicated to this technology, but the engineering community has not been looking for energy anomalies associated with pseudosparks. If they look, they'll likely see the charge cluster (EV) phenomena discovered by Ken Shoulders.

Ed Gray hailed his invention before stockholders, 1976.

EMS --Electronic Power That Could Change The World's Economic Power Picture

59. Edwin Gray. In 1976 Edwin Gray won the prestigious, inventor of the year award for his pulsed capacitor discharge engine. He invented an electric motor that exhibited large torque, consumed little power, and ran cool. In early press releases Gray naively announced that he made a fuelless motor, which later caused him to run afoul with the Securities and Exchange Commission. Gray's prototype motors were confiscated, and he was tried for fraud. Later, Gray moved his lab from Los Angeles to Council, Idaho and Sparks, Nevada to escape harassment. In 1989 Gray was found dead in front of his Nevada lab; his death shrouded in mystery. (Lindemann, 2001)

Gray

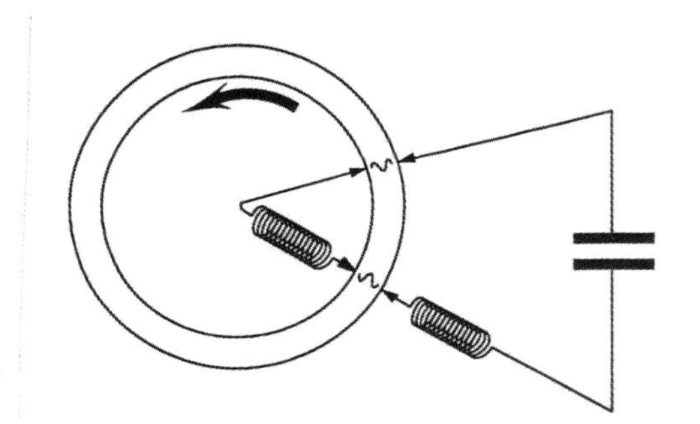

Pulsed Capacitor Discharge
Electric Engine
U.S. Patent 3,890,548

60. Gray's motor. Gray's powered his motor from circuitry, which launched electrical pulses that exhibited a cold current characteristic. Each pulse propagated through a stator electromagnet, jumped across a spark gap to energize a rotor electromagnet, returned back across another gap to excite a different pair of stator-rotor electromagnets, and then was redirected back to the battery to recharge it (Gray, 1976). Magnetic repulsion between the stator and rotor magnets propelled the rotor. To a trained electrical engineer this approach to charging the magnets appears absurd since the sparking would seem to waste energy. Yet here was a motor that was driven by unusual "cold current" pulses, which exhibited spectacular efficiency.

Scalar Compression Tube

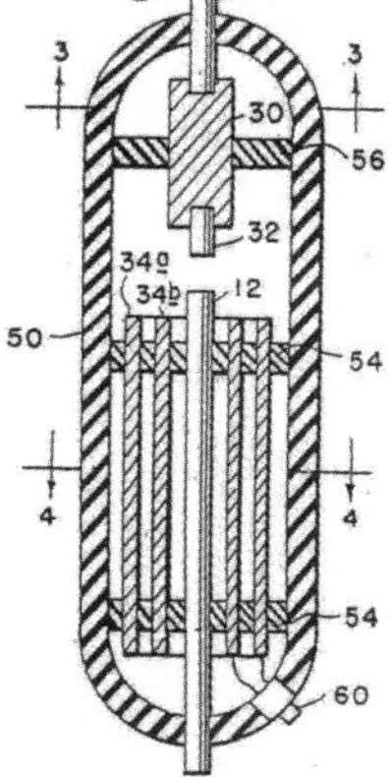

E.V. Gray, U.S. Patent 4,661,747 (1987)
U.S. Patent 4,595,975 (1986)

61. Gray's tube. In 1986 (and again in 1987) Gray patented a circuit and a tube. The tube is the critical element and it exhibits characteristics similar to hollow cathode switches and the tubes of Correa and Moray. The thin anode, down the central axis of the tube, contains a spark gap. The anode is surrounded by a cylindrical, double grid cathode with the two grids electrically shorted together. The double grid behaves like a hollow cathode and contains glow plasma. This is the key component; it is noteworthy that the 1987 patent is identical to the 1986 patent except for one claim that stressed the importance of the double-wall grid cathode. The cold current pulses originate from the glow plasma within this cathode.

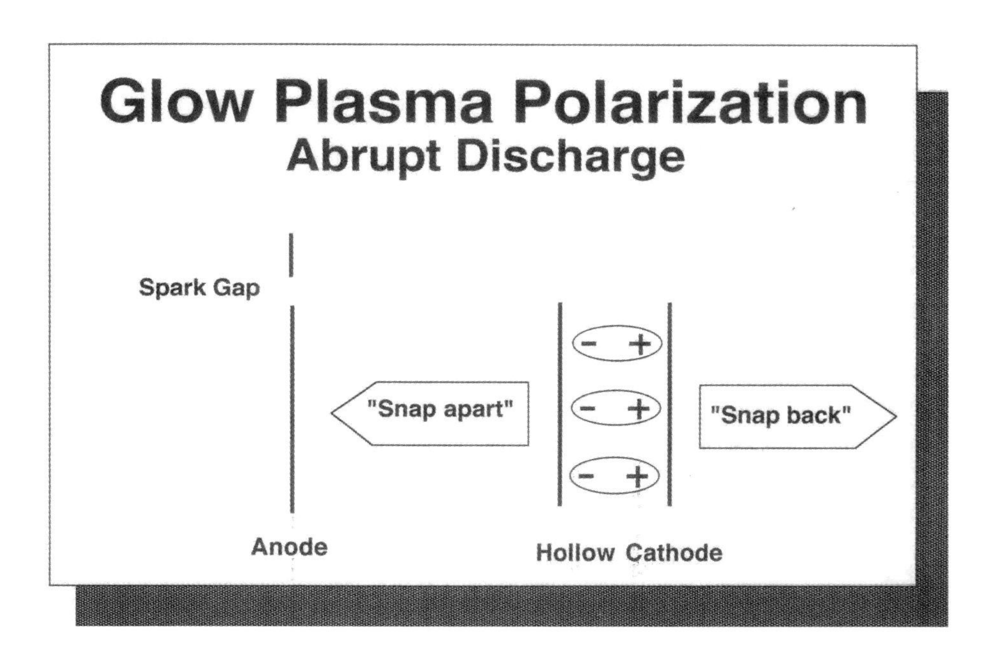
Glow Plasma Polarization
Abrupt Discharge
Spark Gap
"Snap apart"
"Snap back"
Anode
Hollow Cathode

62. Glow plasma polarization. Gray's tube can repeatedly trigger abrupt ion movement in the glow plasma within the double-grid cathode. The plasma is polarized by high voltage (2 KV), which is maintained between the anode and the cathode by a capacitor. The anode spark gap, which is triggered by a control circuit, serves three purposes: 1) It abruptly reduces the anode voltage causing the polarized plasma to "snap back," where the plasma electrons surge outwardly and the ions jerk inwardly. 2) The spark induces further ionization of the glow plasma by photon bombardment. 3) The spark partially ionizes the gas in the gap between the anode and the cathode activating it to the threshold of breakdown, a state known as the "Townsend region" (Lagarkov, 1994). The Townsend region can support a polarization wave with little electron conduction (no arc). It appears to be at the peak of the negative resistance zone as identified by Correa. The "snap back" release of the polarized plasma is like snapping a stretched rubber band. The induced abrupt inward surge of the ions is the critical activator of the vacuum polarization pulse, the mode that produces the cold current effect and manifests an efficient energy gain.

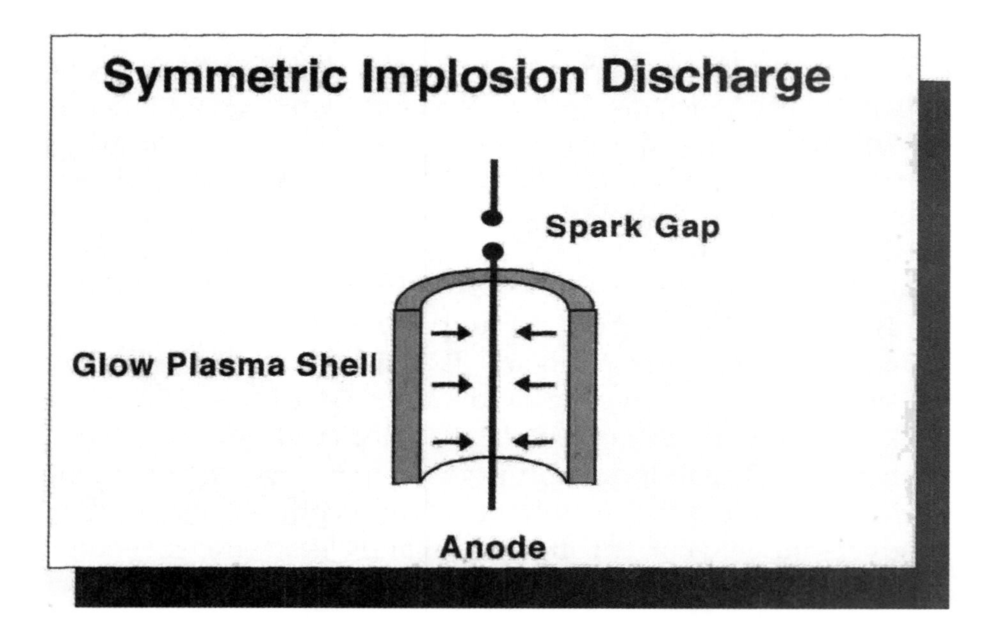
Symmetric Implosion Discharge
Spark Gap
Glow Plasma Shell
Anode

63. **Symmetric implosion.** Note that the cylindrical geometry of the Gray's cathode grid supports a radial, inward ion surge. This could manifest a "scalar compression" pulse where a positive polarization wave is symmetrically directed inwardly to surround the anode. The phrase "scalar" means scalar potential, which results when opposite electric field vectors cancel. Because of symmetry, the radial electric field vectors from the ions are in perfect opposition, which creates a spiking, scalar potential transient when they surge inwardly. If such a transient also cohered the vacuum energy, it could further energize the polarization wave. The wave would then enter the circuit on the conduction path from the anode, and it might be the basis for an even more energetic, cold current pulse. At the same instant the ions surge inward, an electron polarization pulse is directed outward from the grid onto the circuit conductors attached thereto. The "snap back" polarization event retains most of the glow plasma particles within the double wall grid. The event requires just a small, abrupt release of charge from the high voltage capacitor, produces minimal electron conduction current, and provides an efficient way to trigger an ion surge and polarization pulse. The glow plasma membrane acts like a drumhead to launch the polarization wave. The configuration appears to offer researchers a simple straightforward approach to tap the zero-point energy.

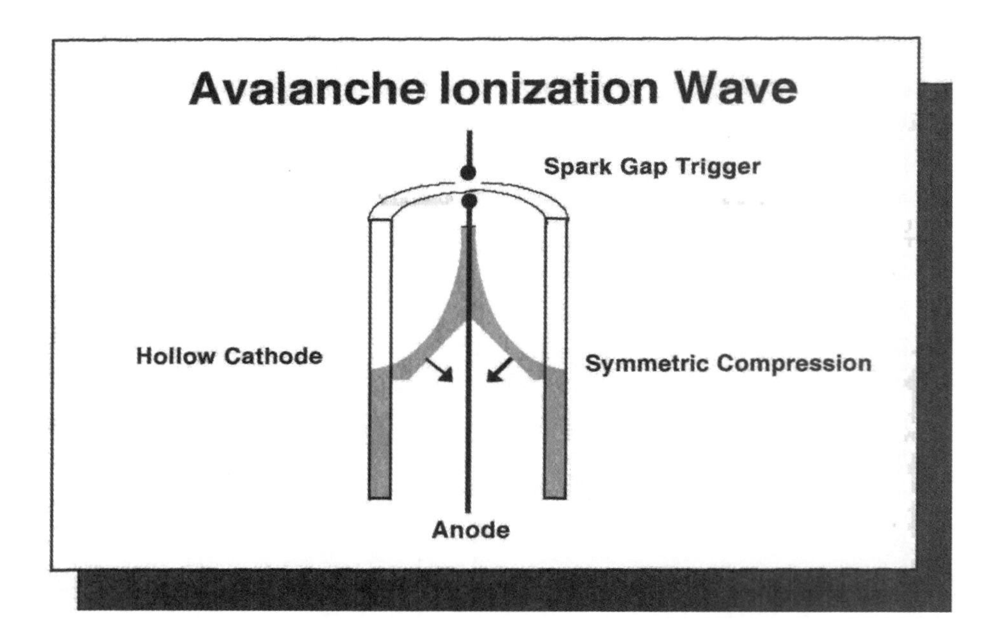
Avalanche Ionization Wave
Spark Gap Trigger
Hollow Cathode
Symmetric Compression
Anode

64. Avalanche discharge. Gray's tube can manifest another discharge mode where the polarized glow plasma "snaps apart" causing a complete discharge event. This behavior is similar to the pseudospark switch, where the anode spark gap acts as the trigger causing a photo-ionization, avalanche breakdown between the cathode and the anode. In this mode the contents of the capacitor is completely discharged causing a plasma compression to occur toward the anode. This event would likewise surge ions (following the electrons) symmetrically inward toward the central electrode, and could thus activate coherence in the zero-point energy as well. It appears that Gray did not want to operate in this mode because in the patent he described an outwardly directed electron pulse, and his system included a circuit breaker component to help protect against a big capacitor discharge event. Nonetheless, since this mode involves abrupt ion motion, it might likewise activate vacuum energy.

Plasma Focus

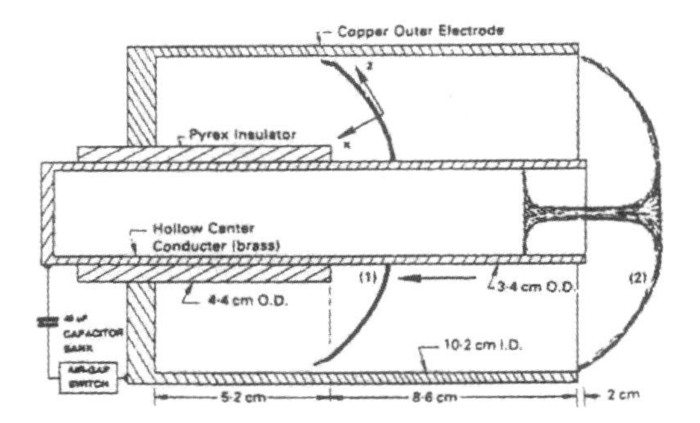

Eric Lerner, www.focusfusion.org

65. **Plasma focus device.** It is interesting to note that the plasma compression mode of Gray's tube has similarities to the plasma focus device. The device was originally invented in the late 1950's to create fusion, and was also known as the zeta pinch device. In the 1960's Bostick and Shoulders collaborated on some experimental studies, and they observed some interesting anomalies including a micron size bore hole that was created right up the central axis of the central anode. Shoulder's later learned that charge clusters (EV) caused this bore hole. Energetic events were observed, but the strong x-ray emissions were attributed to electron bombardment onto the anode. Hot fusion scientists abandoned the zeta pinch device in favor of the tokamak. However, there has been a recent resurgence of interest in the plasma focus device because recent experiments (Lerner, 2002) have shown that the energetic activity is coming directly from the plasma and not from anode bombardment.

Plasma Focus

1 Billion Degrees K

Outputs electricity

No neutrons

No radioactive waste

$$H^{1}_{1} + B^{11}_{5} = 3\,He^{4}_{2}$$

Eric Lerner, www.focusfusion.org

66. Focus Fusion Advantages. Eric Lerner (2002) and his collaborators in the Focus Fusion Society experiment with small size devices. A megawatt reactor could be housed within a garage. The reactor produces extraordinary high temperatures sufficient to ignite hydrogen-boron fusion, a reaction that yields only helium as a by-product. The reaction is thus completely free of radioactive contamination. Moreover, their proposed system can output energy directly as electricity instead of heat, which offers a considerably higher efficiency for practical power generation than other approaches to fusion.

Focus Fusion Plasmoid

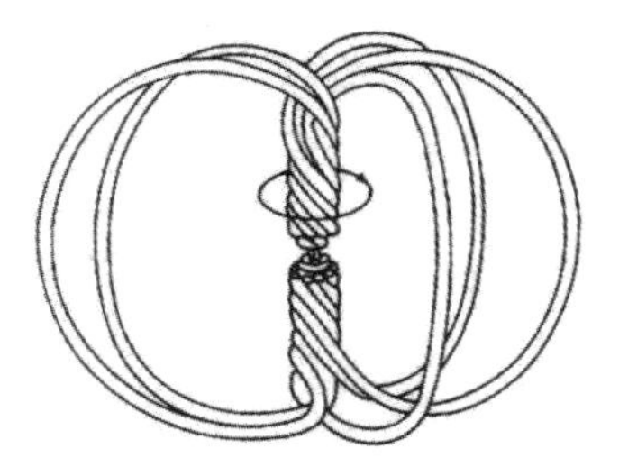

Emits ion and electron beams

Eric Lerner, www.focusfusion.org

67. Focus Fusion Plasmoid. Experiments have shown that the energetic events are coming directly from a generated plasmoid (Bostick, 1957), a coherent plasma form resembling ball lightning. As the plasmoid decays it emits electron and ion beams in opposite directions. The focus fusion system converts these emissions directly into electrical pulses. For the most part, the plasma research community hypothesizes fusion as the source of energy, but there could be a surprise in store for the research teams exploring plasma focus: Since they are doing the right activation to fulfill the zero-point energy hypothesis, where the plasmoid itself could coherently couple with the ZPE, they might discover excessive output energy using just inert gas without inducing any fusion.

The Free Energy Secrets of Cold Electricity
Peter Lindemann, DSc

68. ***Secrets of Cold Electricity.*** Peter Lindemann (2001) has made a significant contribution to the new energy community by generously publishing his research and analysis of the invention of Edwin Gray. Because of his book, there are many hobbyists replicating Gray's tube and observing the energy anomalies. Gray's tube is an excellent starting point for any researcher who wishes to see successes quickly. Any high voltage technique can be used to create a band of glow plasma within a double-wall grid, and there are many methods that could be utilized to abruptly "snap" the plasma polarization. A new generation of energy inventors might arise from just applying the principle of abruptly surging ion movement in glow plasma.

Testatika Machine

Paul Baumann

P. Lindemann, The Free Energy Secrets of Cold Electricity

www.free-energy.cc

69. Swiss ML converter. The famous Swiss ML converter, ("testatika" device), invented by Paul Baumann, was a self-running energy machine that directly output kilowatts of electricity (Matthey, 1985). It was used to provide electrical power to a small religious community. Since they felt mankind was not yet ready to receive the discovery of this energy, they withheld technical information from the numerous visitors and witnesses. Counter-rotating acrylic disks with metal segments like a Wimshurst machine appeared to power the device. The disks produced high electrostatic voltage and colorful swirling plasma. It directed the voltage onto two sealed cylindrical chambers described as "Layden jars." Although witnesses were free to examine the rotating disks, the contents of the "jars" were kept secret. Lindemann suggests that the cylindrical chambers were essentially operating like Gray's tube. A band of glow plasma is induced in a double wall grid around the chamber's circumference, and pulsing activity triggers ion motion. The resulting polarization pulses can be stepped down in voltage and rectified onto output capacitors by standard electrical engineering techniques. If experiments confirm that ion surges do indeed couple ZPE, many historic "free-energy" inventions might be explained from this hypothesis.

Moray Principle

1. Oscillate ions in a glow plasma.

2. Use radioactivity as a catalyst to make the plasma.

3. Tune circuit elements to resonance.

70. Moray's principle. T.H. Moray stressed the importance of ion movement and oscillations, and invented a full energy system based on this principle. Paul Brown showed how plasma oscillations could be maintained easily at low threshold voltages by use of weakly radioactive materials. Paulo and Alexandra Correa stressed the importance of working with the precursor discharge in glow plasma, and Ken Shoulders has discovered that charge clusters can arise from such precursor pulses. Edwin Gray appears to mimic just the precursor without the discharge by abruptly surging ion movement in a band of glow plasma. Often the inventors stimulated and further ionized their plasma by a small electric discharge, but were careful to avoid switching the system into a full electric arc. The theories of vacuum polarization in the zero-point energy support the hypothesis that abrupt synchronous ion movement might yield a net energy gain. A study of the tubes in Moray's patent shows he was a master of corona engineering, and perhaps his free energy machine was the most sophisticated ever invented in the history of the field. T.H. Moray can certainly be credited for the discovery that ion movement is a key activator for manifesting anomalous energy.

Engineering Principles

1. Abrupt motion of glow plasma nuclei
   Polarize / Discharge:
   "Snap back"
   "Snap apart"

2. Shape glow plasma
   Mobius
   Caduceus
   Symmetric compression

3. Bucking EM fields

4. Counter-rotation / Vortical Forms

Stimulate Glow Plasma

1. Abrupt Electric Pulse

2. Bucking EM Fields

3. Counter Rotating, EM Fields

71. Engineering principles (King, 2001). The principle of abrupt ion motion is just one of the engineering principles for interacting with and cohering the zero-point energy. The principle of counter-rotation is based on conserving angular momentum to mimic pair production: Coherent forms arising from the vacuum energy seem to occur in pairs that conserve charge and spin. Another principle is to create abrupt transients of opposing electric fields or magnetic fields. These create a dynamic "scalar" coherence in the ZPE that could couple further energy into the glow plasma. Combining all the principles can lead to many new energy inventions. It is hoped that inventors try the ideas and share what they learn with the scientific and engineering communities. Working together, we can offer humanity a wonderful future founded on a new source of energy.

Acknowledgement
The author wishes to thank Tom Valone for his encouragement and his generous sharing of resource material.

SUPPRESSION

Suppression

1. Academic (paradigm violation)
2. Block funding
3. Block patents
4. Litigation
5. Threats
6. Frame with crime
7. Property destruction
8. Assassination

Author's Note: I typically show the following slides only if the audience brings up the topic of suppression. I did not originally plan to include them in this book. However, the recent brutal murder of Dr. Eugene Mallove, editor of Infinite Energy magazine and popular lecturer, motivated me to include them.

1. Suppression Tactics. New energy devices, whose source is unknown or (perhaps) the zero-point energy, typically are suppressed especially if they appear successful enough to become commercial products. The academic community declares such devices as frauds because they refuse to recognize that the zero-point energy might be a source. Those few professors who dare to acknowledge the possibility are ridiculed and shunned, much like what occurred during the cold fusion debates. Funding is blocked and patents refused by accusing the device of being a "perpetual motion machine," which violates conservation of energy.

Patents are also blocked by issuing a secrecy order. The patent application is stamped top secret, and the work is declared to be a threat to national security. All research must cease and those involved can be prosecuted for treason if they discuss the topic. The government pays no remuneration and those who invested lose everything.

Another tactic that plagues small business is to use incessant litigation so that capital funding is depleted in legal fees. Accusations of irregularities in fund raising are common, and even true, especially if the inventor is unaware of the subtleties of business law. Upon accusation, the prosecution can confiscate records, computers, and technical equipment. Even if the inventor eventually wins the case, the project is effectively shut down.

Threats of physical harm to the inventor or his family is a frequent tactic. Late night anonymous phone calls are typical and are designed to scare the inventor into quitting. Often this tactic works because few inventors have the resources to defend themselves.

Another tactic is to frame the inventor with a crime. Sometimes false witnesses appear to testify. Sometimes criminal evidence such as illegal drugs are planted in the inventor's home.

Historically property destruction such as ransacking or destroying the

Levels of Suppression

1. Academic: Violates paradigm
2. Business: Eliminate competition
3. Black Operations: Security issues

laboratory or fire bombing a vehicle has occurred.

Assassination occurs when the inventor refuses to quit in face of the threats. I personally know four colleagues, fellow speakers at past energy conferences, who have been murdered (Stephen Marinov, Stan Meyers, Paul Brown, Eugene Mallove).

2. Levels of Suppression. Who is behind the suppression? What are their motives? There appears to be three levels of suppression:

Level one is from the academic community. They are motivated by the need to be right. They are protecting the current paradigm that space is effectively devoid of useable energy. Thomas Kuhn's, The Structure of Scientific Revolution, shows that history is replete with examples of the scientific community refusing to explore or acknowledge new discoveries that would shake their world view. Their suppression tactics are usually benign, and include ridicule, peer-review rejection, shunning, accusations of fraud, and most often simply ignoring the field of research.

Level two is from industry. They could believe that new energy devices exist which could dramatically shift profits from today's entrenched energy industries. They are motivated to protect their businesses and could hire thugs or operatives to apply ruthless suppression. History has numerous examples, and the motivation is to maximize profits from the status quo.

Level three is from the black operations community of the military-industrial complex. They believe the new energy discovery is possible, and that such a discovery might be dangerous. Their motivation is protection and security, and such a motivation is honorable.

3. Disclosure Project. Dr. Steven Greer has started an initiative that might offer a solution to suppression. Greer has made the point with his acquaintances in the government that the elected officials do not control the secret, black operations of the military-industrial complex. Many citizens agree that these projects should be under control of our elected government. These citizens include numerous employees participating in black projects. They are willing to testify to congress about the discoveries of new energy and propulsion technologies if they are granted immunity from their secrecy oaths. Some have already testified that not only have new technologies been developed, but more astonishingly, secret contact with extraterrestrial intel-

TOP SECRET
RESTRICTED
PRE-PUBLICATION
RUSH EDITION
DISCLOSURE
MILITARY AND GOVERNMENT
WITNESSES REVEAL
THE GREATEST SECRETS
IN MODERN HISTORY
STEVEN M. GREER M.D.
www.DisclosureProject.org

ligences has occurred. This contact has been the source of the new technologies. Greer's point is that if this is true, then such contact should be with the elected government not black operations.

The implications of the Disclosure Project are astounding and hopeful. It appears we have been guided to make the new energy discovery. Inventors often tell how their inspiration came from visions, dreams, revelation, spiritual contact, UFO encounters, or synchronicity. The energy discovery could effectively be used to meet mankind's physical needs including cleaning up the environment, yet the discovery is double edged and potentially dangerous. It seems that we would not be gifted with such knowledge unless there would be help and guidance for its safe use. It just might be that a widespread discovery of zero-point energy technology could trigger formal, public contact with advanced spiritual intelligences who are here to guide mankind to use it wisely. Perhaps they are patiently waiting for us to invite them.

It appears that the wise use of potentially limitless energy requires a shift of consciousness. Much spiritual literature describes a mature consciousness that logically recognizes and emotionally feels that all life is sacred and interconnected. Peter Russell in his book, The Global Brain, offers the analogy that human beings are like brain cells for a planetary consciousness "Gaia" that has not yet awakened. Perhaps a minimal consciousness shift would involve the ability to feel the other's pain whenever we hurt someone. The state of mind expressed from the first person is "I am all men." A generalization at the planetary or universal consciousness level is "We are all races." The point is that we no longer slay one another or go to war because we all recognize "I am the other person; what I do to another I do to myself." Such a state of mind has been described as a "millennium consciousness." It just might be that the widespread discovery of zero-point energy technology could trigger a series of events culminating in a consciousness transformation of the human race. We live in exciting times.

PATENT OF T. HENRY MORAY

Feb. 1, 1949. T. H. MORAY 2,460,707

ELECTROTHERAPEUTIC APPARATUS

Filed April 30, 1943 3 Sheets-Sheet 1

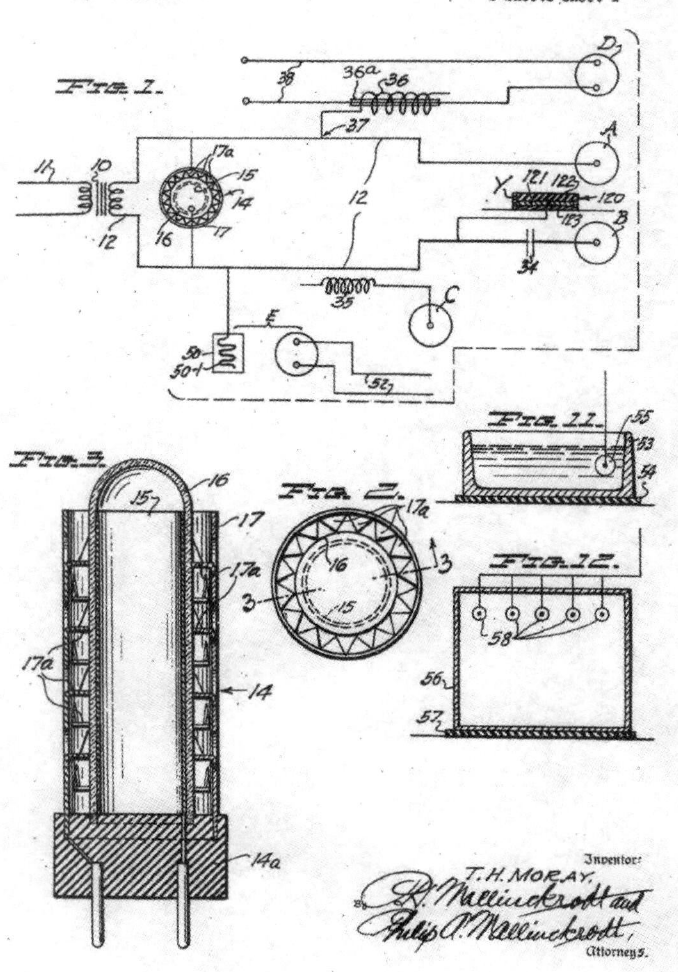

Feb. 1, 1949. T. H. MORAY 2,460,707

ELECTROTHERAPEUTIC APPARATUS

Filed April 30, 1943 3 Sheets-Sheet 2

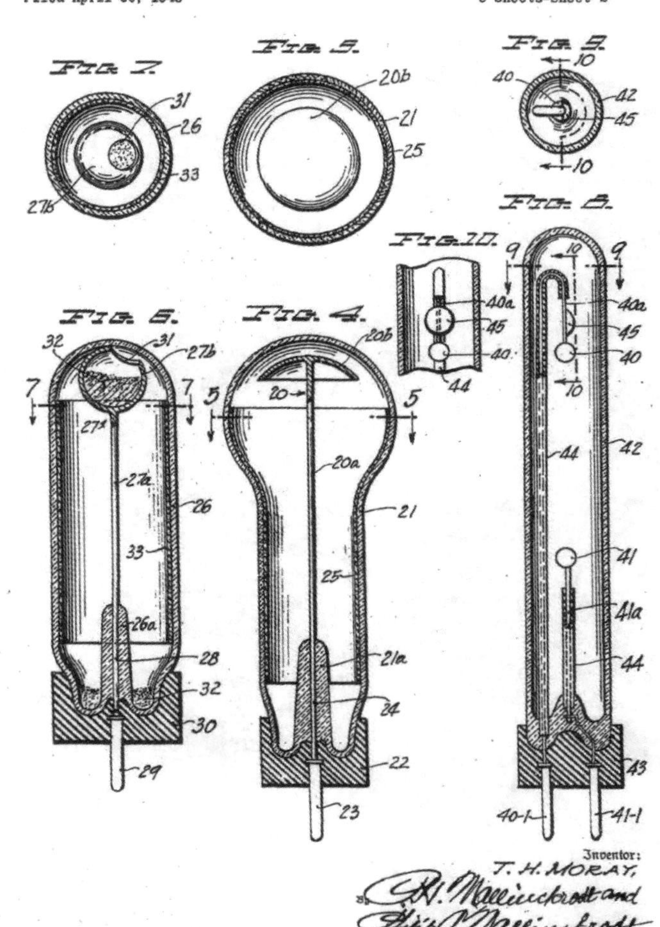

Patented Feb. 1, 1949 **2,460,707**

UNITED STATES PATENT OFFICE

2,460,707

ELECTROTHERAPEUTIC APPARATUS

Thomas H. Moray, Salt Lake City, Utah

Application April 30, 1943, Serial No. 485,112

8 Claims. (Cl. 128—421)

1

This invention relates to electrotherapeutic apparatus, and to methods of applying electrical, radioactive, and other radiant phenomena therapeutically.

The invention is primarily concerned with the use of high potential, high frequency electricity though not necessarily limited thereto, in conjunction with radioactive and other types of electronic and radiation phenomena, for therapeutic purposes.

Among the objects of the invention are the following:

First.—To render highly effective, from a therapeutic standpoint, radioactive and other types of electronic and radiation phenomena, and, likewise, to render highly effective, from a therapeutic standpoint, high potential, high frequency electricity.

Second.—To augment the therapeutic effect of radioactive and other types of electronic and radiation phenomena by the conjoint use of high potential, high frequency electricity, and, conversely, to augment the therapeutic effect of high frequency, high potential electricity by the conjoint use of radioactive and other types of electronic and radiation phenomena.

Third.—To accomplish the above without danger of burning or of otherwise harming the patient.

Fourth.—To provide apparatus for accomplishing the above, which is relatively simple in construction and operation and relatively inexpensive to produce and operate.

Fifth.—To provide novel electronic and radioactive devices especially adapted for use in conjunction with high potential, high frequency electrical therapy.

I have found that, by enveloping a patient in a high potential, high frequency electrical field in such a manner that no closed circuit is completed through his body, radioactive and other electronic and radiation phenomena can be used therapeutically with considerably greater effectiveness than if used alone. The exact reason for this is not known, nor is it known definitely which, the electric field or the radioactive phenomena, acts upon the other to produce the advantageous results. It is thought, however, that the electric

2

of malignant tumors, arthritis, sinus infections, and various other diseased conditions.

The invention contemplates the use, in therapeutics, of high potential, high frequency electricity to produce diversified forms of radiant energy, such forms being those which have been found best suited, individually, to benefit various human ailments. In accomplishing this purpose, several special discharge tubes have been developed to serve as treatment electrodes, by means of which correspondingly different curative results are obtained. Throughout the practice of the invention, a prime consideration is that only one terminal of any particular circuit shall be in contact with a patient's body at one time, so there will be no flow of current through a closed circuit of which the patient's body is a part. Such a terminal, too, is usually non-heat producing, so there is no danger of burning. In cases where there is a tendency for a tube to produce X-rays or other injurious rays, these are filtered out.

The present application constitutes a continuation in part of a copending application filed by me November 15, 1940, which bears Serial No. 365,798 and is entitled "Method of and device for the therapeutic application of electric currents and rays," and which has now become abandoned.

In the accompanying drawings, which illustrate several embodiments of apparatus preferred for carrying the method of the invention into practice:

Fig. 1 represents a wiring diagram of a preferred embodiment of apparatus for carrying out the method of the invention in general therapeutic work, several independent treatment stations being provided;

Fig. 2, a top plan view of the novel corona regulator of Fig. 1, employed in the circuit to control and adjust the current and as a governor to safeguard the transformer;

Fig. 3, a vertical section taken on the line 3—3, Fig. 2;

Fig. 4, a vertical section taken centrally through one novel type of discharge tube used as a treatment electrode in the apparatus of Fig. 1;

Fig. 5, a horizontal section taken on the line

Feb. 1, 1949. T. H. MORAY **2,460,707**

ELECTROTHERAPEUTIC APPARATUS

Filed April 30, 1943 3 Sheets-Sheet 3

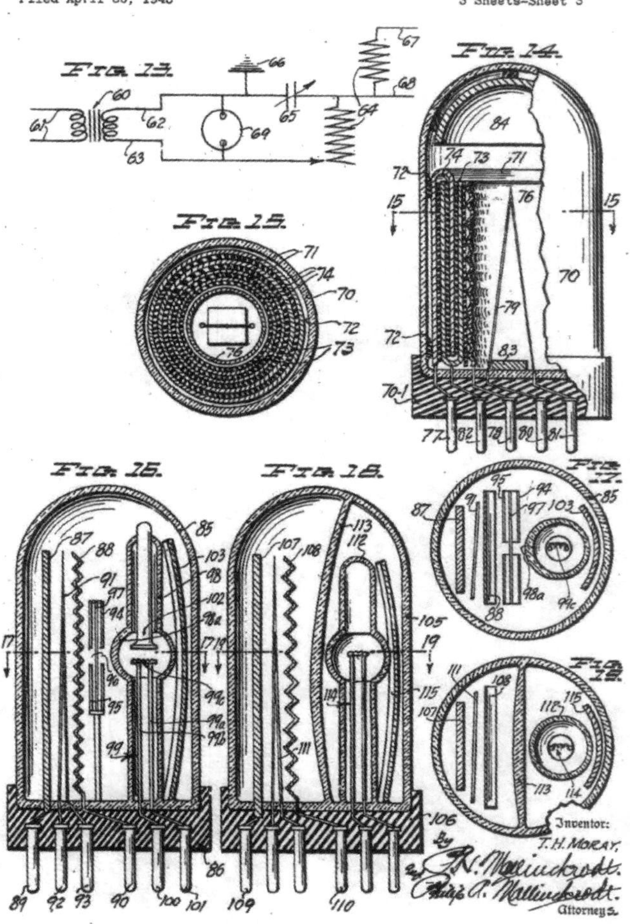

a novel discharge tube used as a treating device in the apparatus of Fig. 1;

Fig. 9, a horizontal section taken on the line 9—9, Fig. 8;

Fig. 10, a fragmentary vertical section taken on the line 10—10, Figs. 8 and 9;

Fig. 11, a fragmentary view in vertical section, and drawn to a reduced scale, of a tub bath capable of use as a treatment station in the apparatus of Fig. 1;

Fig. 12, a view similar to that of Fig. 11, but showing a shower or vapor bath arrangement for the same purpose;

Fig. 13, a wiring diagram similar to that illustrated in Fig. 1, but fragmentary in nature, and of a somewhat different embodiment of apparatus;

Fig. 14, an elevation, partly in central vertical section, of a novel tube used in the apparatus of Fig. 13 in place of the corona regulator of Figs. 2 and 3;

Fig. 15, a top plan view, partly in horizontal section on the line 15—15, Fig. 14, of the tube of Fig. 14;

Fig. 16, a vertical section of another novel tube which may be used in place of the tube of Figs. 14 and 15;

Fig. 17, a vertical section taken on the line 17—17 of Fig. 16;

Fig. 18, a top plan view of still another novel tube which may be used in place of the tubes of Figs. 14 and 15 and of Figs. 16 and 17; and

Fig. 19, a vertical section taken on the line 19—19 of Fig. 18.

In accordance with the invention, provision is made for enveloping the patient in a high potential and, in certain instances, a high frequency electric field, and for applying to the patient, while so enveloped in the electric field, radiations and emanations having therapeutic value.

The apparatus of Fig. 1 is capable of administering various specific kinds of treatment, pursuant to the invention, at the several treatment stations provided. The treatment stations are indicated A, B, C, D, and E, respectively.

For supplying the high potential electric field, a suitable transformer is employed. This may be of any type capable of delivering high potential electricity, say from 10,000 to 30,000 volts. It is preferred, however, to utilize a conventional double magnetic circuit type of transformer, indicated at 10 in Fig. 1, having adjustable, laminated, magnetic shunts (not shown), the transformer being connected across an ordinary power line 11 charged with the customary 115 v. The output lines 12 from this transformer advantageously extend to the treatment stations A and B, respectively. The first secondary of the transformer 10 is preferably direct connected to the second secondary thereof. It is noted that this high potential electricity may be applied, without causing injury, direct to a patient who is not grounded. However, in order to safeguard the transformer 10 from damage by sparking across its output terminals, and to render the high potential electricity more suitable for therapeutic purposes, which is believed to include the automatic changing of the frequency to an extent which depends upon electrical characteristics of the patient's body, a governor or control device 14 is shunted across the leads 12.

This governor or control device 14 is a sparking condenser of high capacity embodying a multitude of spark gaps. A preferred embodiment of this governor or control device 14 is illustrated in detail in Figs. 2 and 3.

As illustrated, the device comprises a cylindrical, electrically conductive plate 15 surrounded by a cylindrical dielectric 16. An outer cylindrical and electrically conductive element 17 surrounds the dielectric 16 exteriorly. It is provided with a multitude (for example, 250) of inwardly extending prongs 17a, which are advantageously formed by stamping out, and inturning, triangular portions of the electrically conductive element 17. The internal plate 15 preferably contacts the interior surface of the dielectric 16, but, in any event, should lie closely adjacent thereto. Likewise, the tips of the prongs 17a preferably contact the outer surface of the dielectric. The several elements are advantageously mounted in a plug-in base 14a, which is adapted to mate with a suitable receiving socket (not shown) carrying the required electrical connections. The internal plate 15 connects with one of the electric lines 12, while the external element 17 connects with the other electric line 12, as shown diagrammatically in Fig. 1.

It is preferable that the dielectric 16 be in the form of a closed tube or envelope, as shown, and be exhausted to vacuum condition. The multitude of sparking prongs 17a product a brush discharge.

Where the dielectric 16 is not a closed tube or envelope, it is preferred that it be of quartz.

The treatment station A is a discharge tube of a novel type, exemplified by the tubes illustrated in detail in Figs. 4 and 5 and Figs. 6 and 7. Either tube is plugged into the circuit of Fig. 1 at a suitably provided, single-terminal outlet. High potential electricity is, therefore, fed directly into the tube, which serves as an electrode. The tube also embodies radioactive material, which supplies radioactive emanations to the patient simultaneously with the electrical discharge.

As illustrated in Figs. 4 and 5, the tube or electrode may comprise an electrically conductive discharge element 20, having a supporting stem 20a and a major discharge cap or head 20b, which is preferably in the form of a thin, convex-concave plate. The head 20b may be spot welded to the end of the stem 20a.

The discharge element 20 is enclosed within a tube 21 of dielectric material, preferably glass, the stem 20a being fixed in the fused tongue portion 21a of the tube. The tube or shell 21 is fitted into an insulating base 22, provided with a single plug-in terminal 23, and an electrical connector 24 extends from the terminal 23 to the stem 20a.

The inside surfaces of the side walls of the tube or shell 21 are coated with a radioactive material, as at 25. The coating is conveniently made from uranium salts or powdered carnotite or other radioactive ore. The ends of the tube or shell are left uncoated.

Air is evacuated from the tube 21, and a small quantity of mercury introduced. The mercury is preferably triple-distilled to insure great purity. It is preferred that argon or like inert gas be also introduced.

Since the tube just described is plugged into the circuit of Fig. 1, the discharge element or cathode 20 is charged with high potential electricity, and, in its capacity of a treatment station in the apparatus of Fig. 1, serves as an electrode to similarly charge the patient. The patient is insulated from the ground, and the tube is applied directly to the afflicted part of his body, preferably in close contact with the body.

Because of the construction of the tube, radiation of a radioactive nature is also directed against the patient through the uncoated top end of the tube. This radiation has been found to differ somewhat from the radioactive emanations discharging from the side walls of the tube, and is thought to comprise rays lying close to X-rays on the radiation spectrum. These rays appear to have a definite healing value, and to lack the injurious nature of X-rays. Where a predominantly radioactive emanation treatment is desired, the side walls of the tube are placed against the body of the patient.

Best results are obtained when the discharge element or cathode **20** is made of an alloy metal compounded from copper, lead, sulphur, and, if desired, aluminum. The relative percentages of the several ingredients may vary considerably, but a satisfactory mixture comprises 5.0% copper, 55.0% lead, 30.0% sulphur, and 10.0% aluminum. Should aluminum not be used, the difference may be made up by additional copper.

In preparing the alloy, the copper and aluminum are heated to a molten state, after which the sulphur is added while stirring the mixture. After cooling, the mass is again melted, and the lead, in a molten state, is mixed with it, the molten mass being thoroughly stirred. This new mass is then cooled, being later reheated, and, while hot, rolled to make it ductile, so it can be shaped into the desired forms.

The discharge tube or electrode of Figs. 6 and 7 is similar to that of Figs. 4 and 5, having an enclosing tube or shell **26** which is evacuated. A cathode discharge element **27** is positioned within the shell, being fixed in the tongue portion **26a**. A conductor **28** connects the stem **27a** of the element **27** with a plug-in terminal **29**, which extends outwardly of the base **30**. The cap or head **27b** of the element **27** differs from the cap or head **20b** of the electrode of Figs. 4 and 5, in that it is spherical in form and hollow. It has an opening **31** formed at its top, contiguous with the top inside surface of the tube **26**. A quantity **32** of radioactive material, which may be the same as used for the coating **25** of the electrode of Figs. 4 and 5, is introduced into the tube or shell **26**, along with a relatively small quantity of mercury, before the tube is sealed tight. Such material **32** is preferably powdered or granulated, and is shaken into the hollow of the head **27b** through the opening **31** before any given treatment is commenced. The mercury is provided primarily as a getter, and does no harm if shaken into the head **27b** along with the radioactive substance. The mercury also tends to produce a vapor in the tube, which aids in the operation thereof. As in the case of the electrode tube of Figs. 4 and 5, this tube may have a radioactive coating **33** covering the inner surfaces of its side walls.

The treatment station B of Fig. 1 differs from the treatment station A only in the fact that a condenser **34** is interposed in the electric supply line **12**.

The treatment station C of Fig. 1 differs from the stations A and B only in the fact that the high potential electricity is supplied from the supply line **12** through an inductance **35**.

The treatment station D utilizes a germicidal discharge tube, a preferred form of which is illustrated in detail in Figs. 8, 9, and 10. The high potential electricity is taken by induction from the particular supply line **12** concerned. For this purpose, an induction coil **36** is provided, tapping the line **12** at **37**. A pair of leads **38** from an ordinary 115 v. supply source extend to a plug-in socket connection for the germicidal tube, one of the leads passing through a glass tube **36a**, Fig. 1, which is disposed within and extends along the length of the induction coil **36**. Thus, high potential electricity is impressed, by induction, upon the ordinary current flowing through the particular lead **38** concerned.

The germicidal discharge tube of Figs. 8, 9, and 10 has a pair of discharge terminals **40** and **41**, respectively, positioned in an evacuated tube or envelope **42**, and electrically connected with plug-in terminals **40—1** and **41—1**, respectively, by means of stems **40a** and **41a**, respectively. The tube or envelope **42** and plug-in terminals are mounted in a conventional base **43**. It is preferred that insulating material **44**, such as a ceramic sleeve, cover the major portions of the stems **40a** and **41a**. A piece of lithium metal **45**, see particularly Fig. 10, is advantageously secured to the stem **40a** adjacent the discharge terminal **40** to act as a getter. It may, however, be placed at any other convenient location in the tube. It is preferred that the discharge terminals **40** and **41** be formed of the special alloy previously described. Argon or other suitable inert gas is preferably injected into the tube or envelope **42**, as is, also, a small quantity of mercury. The mercury, by vaporizing, aids electrical arcing between the discharge terminals. As will be noted, the high potential electricity induced in the one lead **38** will manifest at the upper discharge terminal **40**, and will charge the patient simultaneously with the discharge into his body of germicidal rays from the tube.

The treatment station E embodies the tube of Figs. 8, 9, and 10, as above described, but impresses the high potential electricity directly on the patient instead of passing it first through the tube. For this purpose, a discharge device **50**, in the form of a soft, flexible pad in which a coil **50—1** is embedded, taps one of the high potential electric lines **12**. This pad **50** is wrapped around the patient's body adjacent the afflicted portion thereof, thus charging the patient. Any other electrode capable of charging the patient with high potential electricity may be used in place of the pad **50**. The germicidal tube has its terminals **40—1** and **41—1** plugged into a suitable plug-in socket connected to leads **52** which extend to an ordinary 115 v. source of supply. The high potential electricity with which the patient is charged is induced into the germicidal tube, thereby further activitating the discharge therefrom. A certain beneficial discharge from this germicidal tube will be had by induced activation alone, it being unnecessary, in such instances, to plug the tube into the 115 v. line.

Other types of germicidal and discharge tubes may be used in place of the tube of Figs. 8, 9, and 10, as, for instance, the well known infra-red and ultra-violet lamps, to produce results surpassing those ordinarily attained by the use of such infra-red or ultra-violet lamps apart from the apparatus of the invention.

It should be remembered that the patient is insulated from the ground while being treated at any of the treatment stations of the invention.

Figs. 11 and 12 show how a patient is treated, pursuant to the invention, while immersed in an electrically conductive fluid bath. In Fig. 11, a bath tub **53** is insulated from the ground by a layer of insulation 54. A treatment electrode of the type shown in any of the figure groups 4 and 5, 6 and 7, and 8, 9, and 10 is positioned to charge the fluid of the bath with high potential electricity, as well as to discharge healing radiations and emanations into the patient. The particular electrode illustrated is diagrammatic in form and is designated 55. It may be connected into the circuit of Fig. 1 as shown at any of the treatment stations A, B, C, and D. In Fig. 12, a shower or vapor stall 56 is insulated from the ground by a layer of insulation 57. A plurality of treatment electrodes are designated 58, respectively. These correspond to the treatment electrode 55 of Fig. 11. A water spray or vapor, such as steam, may be admitted to the stall 56 in any well known manner (not shown), thus enveloping the patient during treatment.

Another embodiment of apparatus, pursuant to the invention, is illustrated diagrammatically by the wiring diagram of Fig. 13. While no treatment stations are shown, those provided are identical with the several treatment stations designated A, B, C, D, and E in Fig. 1. The distinction in this embodiment of apparatus resides in the fact that a special generator of high frequency electricity is provided in the system.

A transformer 60 has its input terminals connected across an ordinary 115 v. electric power line 61. Electrical conductors 62 and 63 lead from the respective output terminals of the transformer to a high frequency generator of the Oudin coil type, indicated generally at 64, a variable condenser 65 being interposed in the line 62, and the circuit being grounded at 66. Output conductors 67 and 68, leading from the high frequency generator 64, provide connections for the several treatment stations in the same manner as illustrated in Fig. 1.

The transformer 60 may be any ordinary high voltage type. A governor or control device 69 is shunted across the conductors 62 and 63.

In the illustrated instance, the governor or control device 69 preferably takes the form of a vacuum tube, having the construction shown by Figs. 14 and 15, Figs. 16 and 17, or Figs. 18 and 19. These tubes all possess high capacity, and include elements effecting a brush discharge. They serve, as does the device 14 of Figs. 2 and 3.

The tube of Figs. 14 and 15 embodies an outer shell or envelope 70 of insulating material such as glass, a plastic, or fiber coated with shellac. Inside the shell 70 is a bi-cylindrical element 71 formed of electrically conductive material. Separating element 71 from the enclosing shell 70 are spacers 72 made of rubber, Bakelite, or other insulating material. Inter-fitting with the element 71 is a second electrically conductive, bicylindrical element 73, the two elements being separated by a dielectric 74. Inwardly of the element 73, and separated therefrom by a dielectric 75, is a corrugated, cylindrical element 76. The shell or envelope 70 is secured in an insulating base 70—1, provided with plug-in terminals. One of the terminals, designated 77, is electrically connected with the element 71, while another, designated 78, is electrically connected with the corrugated element 76. These two terminals connect with the conductors 62 and 63, as illustrated in Fig. 13, and the brush discharge takes place at element 76.

Under certain circumstances, it is desirable that the outer shell 70 be made of quartz glass, and that a filament 79 be provided, the filament being heated by connection, through plug-in terminals 80 and 81, with a source of low voltage heating current (not shown). Plug-in terminal 82, which is electrically connected with element 73, may be used instead of or in connection with the terminal 77, since element 73 acts in a manner similar to element 71. A getter 83 of suitable material, and an insulating and reflecting shield 84 may be provided, as shown. While the tube may have either a high or a low vacuum condition, or may be filled with an inert gas, I have also found it advantageous to fill the tube with a moist vapor. The tube acts as an oscillator for electric currents, and has an enormous capacity, a capacity many times that of a condenser of approximately equal size.

The tube of Figs. 16 and 17 comprises an outer shell or envelope 85, which may be made of metal, glass, or fused quartz. This shell is mounted in an insulating base 86. Inside the shell 85 is a metal plate 87, and, spaced apart therefrom, a corrugated metal plate 88. A plug-in terminal 89, which extends from the base, is electrically connected with the plate 87, and a second plug-in terminal 90 is electrically connected with the corrugated plate 88. These terminals are adapted to connect, through a suitable socket, with the electrical conductors 62 and 63 of Fig. 13.

Under certain conditions of use, it is desirable to have other elements in the tube. These are provided, and may be utilized or not as occasion warrants. A filament 91 is disposed between the plates 87 and 88. It is electrically connected with the two plug-in terminals 92 and 93, which are adapted to be connected to a source of low voltage heating current (not shown). A slit screen, comprising shields 94 and 95, with apertures 96 extending therethrough, is disposed adjacent that side of corrugated plate 88 which is remote from plate 87. The apertures 96 are in alignment with each other, and the shields 94 and 95 are made of lead or other material capable of screening off X-rays. Between shields 94 and 95 is a sheet 97 of material which is readily permeable to X-rays. Within the shell 85 there is also mounted a shell or envelope 98 of glass, quartz glass, or similar material, having a portion 98a which is ground like a lens and directed toward the slit screen. This shell 98 really constitutes a tube within a tube. A filament or cathode 99, comprising electrically conductive legs 99a and 99b and an electron-emitting portion 99c, is disposed within the shell 98, plug-in terminals 100 and 101 being electrically connected to the respective legs 99a and 99b. A bombardment element 102 is disposed within the shell 98 opposite the portion 99c of cathode 99. Within the shell 85, but outside the shell 98, is a reflector 103 directed toward the slit screen.

The tube of Figs. 18 and 19 is essentially the same as the tube of Figs. 16 and 17, being equipped with a shell or envelope 105, a base 106, a plate 107, and a corrugated plate 108, the two plates being connected to plug-in terminals 109 and 110, respectively, which are adapted to connect electrically with the conductors 62 and 63 of Fig. 13. There is a filament 111 and an inner shell or envelope 112, but no slit screen. Instead of a lens portion being provided on the inner shell 112, a partition 113 of lens formation is disposed between the inner shell and the corrugated plate 108. It is fused to the walls of the outer shell 105. Within the inner shell 112 is a filament or cathode 114, which corresponds to the similar element 99 of the tube of Figs. 16 and 17. A reflector 115 is directed toward the lens partition 113.

Reverting now to Fig. 1, there is another advantageous way of treating a patient pursuant to the invention. As shown at Y, a foot pedestal 120 may be provided for making the patient a part of a condenser. The pedestal comprises an electrically conductive plate element or electrode 121, connected electrically with one of the high potential lines 12, and covered by an insulating platform 122 upon which the patient rests his feet while being treated at any of the previously described treatment stations A, B, C, D, or E. The electrode 121 and insulating platform 122 are conveniently mounted in a frame 123, which insulates the plate from the ground. The insulating platform 122 is made of a high quality insulating material, such as first grade hard rubber. In certain instances it is desirable that the device be made in other than foot-pedestal form. For instance, it may be of cylindrical formation for use in a bed against any part of the patient's body.

If desired, the patient may be charged with the high potential electricity by direct contact with a metal or electrically conductive electrode in place of the pad 50 of treatment station E, or of the tube electrodes.

The invention has been described in the foregoing with sole reference to its use for therapeutic purposes. It should be noted, however, that inorganic matter may also be treated to advantage pursuant to the method and with the apparatus of the invention. It has been found that metals, for example, lead, have changed physical properties after treatment in accordance with the above. In instances where the invention is not being used therapeutically, it is not always necessary to insulate the subject from the ground.

Whereas this invention is here illustrated and described with respect to particular specific embodiments thereof, it is to be understood that various changes may be made in such specific embodiments and various other embodiments may be utilized by those skilled in the art without departing from the spirit and generic scope of the invention as set forth herein and in the claims which here follow.

Having fully described my invention, what I claim is:

1. Apparatus for applying radiant energy therapeutically, comprising means for producing high potential, high frequency electricity; a high capacity sparking condenser; and a treatment electrode connected in circuit with the foregoing, said treatment electrode including a discharge element adapted to charge the patient with said high potential, high frequency electricity, and radioactive means adapted to discharge radioactive emanations into said charged patient.

2. Apparatus in accordance with claim 2, wherein the sparking condenser is in the form of a vacuum tube of high capacity having mutually spaced capacity elements adapted to produce a corona discharge.

3. Apparatus for applying radiant energy therapeutically, comprising means for producing high potential, high frequency electricity; a high capacity sparking condenser; and a treatment device connected in circuit with the foregoing, said treatment device including discharge means adapted to charge the patient with said high potential, high frequency electricity, and radiating means adapted to discharge radiations into the charged patient.

4. Electrical treatment apparatus, comprising a high capacity sparking condenser; a treatment outlet electrically connected to said condenser; and means for electrically connecting said condenser to a source of high potential electricity.

5. Electrical treatment apparatus, comprising a transformer for producing high potential electricity; a high capacity sparking condenser electrically connected across the high potential output terminals of said transformer; and a treatment outlet electrically connected to said condenser.

6. Electrical treatment apparatus, comprising a transformer for producing high potential electricity; a high capacity sparking condenser electrically connected across the high potential output terminals of said transformer; and a plurality of treatment outlets independently electrically connected to said condenser.

7. Electrical treatment apparatus, comprising a transformer for producing high potential electricity; a treatment electrode electrically connected to one of the output terminals of said transformer; an electrical conductor sheathed by insulation electrically connected to the other of the output terminals of said transformer and disposed adjacent said treatment electrode so the subject to be treated may be placed between and in contact with the two; and a high capacity sparking condenser connected across the said outlet terminals of the transformer.

8. In electrical treatment apparatus equipped with means for the supply of high potential electricity and a treatment electrode, a high capacity sparking condenser electrically connected between the said supply means and the said treatment electrode.

THOMAS H. MORAY.

REFERENCES CITED

The following references are of record in the file of this patent:

UNITED STATES PATENTS

Number	Name	Date
628,351	O'Neill	July 4, 1899
647,687	Topham	Apr. 17, 1900
765,470	Friedlander	July 19, 1904
950,842	Davis	Mar. 1, 1910
1,156,317	Santos et al.	Oct. 12, 1915
1,193,018	Howard	Aug. 1, 1916
1,466,777	Winkelmann	Sept. 4, 1923
1,590,930	Falkenberg	June 29, 1926
1,655,783	Gallois	Jan. 10, 1928
2,010,018	Hodnette	Aug. 6, 1935
2,073,428	Schmid	Mar. 9, 1937
2,126,070	Wappler	Aug. 9, 1938
2,192,638	Dixon	Mar. 5, 1940

OTHER REFERENCES

Tousey; Medical Electricity and Roentgen Rays (1910), pp. 493 to 498. Copy in Division 55.

BIBLIOGRAPHY

References

Adamenko, S.V., A.S. Adamenko, V.I. Vysotskii (2004), "Full-Range Nucleosynthesis in the Laboratory," *Infinite Energy*, vol 8, issue 54, pages 23-30.

Experiments are described where element transmutation is induced by super-compression of a pulsed coherent electron beam focused onto a pure element, metallic target. An hypothesis is offered where plasma nuclei can cluster coherently to produce a super-nucleus, which decays into a variety of elements of unusual isotopes not readily found in nature. Over 5000 experiments have been performed at the Electrodynamics Laboratory, Proton 21, Kiev, Ukraine.

Adams, R.(Jan 1993), "The Adams Pulsed Electric Motor Generator," *Nexus*, pages 31-36.

A professor of electrical engineering claims creating an over unity device.

Aharonov, Y. and D. Bohm (1959), "Significance of Electromagnetic Potentials in the Quantum Theory," *Phys. Rev.* 115(3), page 485; Olariv, S. and I.I. Popescu (1985), "The Quantum Effects of Electromagnetic Fluxes," *Rev. Mod. Phys.* 57(2), pages 339-436.

Akimov, A.E.(1998), "Heuristic Discussion of the Problem of Finding Long Range Interactions, EGS-Concepts," *J. New Energy* 2(3-4), pages 55-80.

Overview of the torsion field research (predominantly in the Soviet Union) includes 177 references. Akimov models the vacuum as a lattice of "phytons," counter-rotating, charged entities sized at the Planck length (10^{-33} cm). Each phyton can polarize in three different ways to manifest 1) electric fields via charge polarization, 2) gravitational fields via oscillating, longitudinal spin polarization, and 3) torsion fields via transverse spin polarization. The gravitational and spin polarizations are hypothesized to support a wave propagation faster than light, and the torsion field supports superluminal links between originally coupled quantum particles to explain EPR "nonlocal" connectivity. Torsion fields can arise from four sources: 1) physical classical spin, 2) the spin of the elementary particles comprising an object, 3) electromagnetic fields, and 4) the geometric form of the object. The spin polarized phyton lattice can also retain a temporary residual torsion field image of a (long standing) stationary object after it is moved.

Alexandersson, O.(1990), *Living Water: Viktor Schauberger and the Secrets of Natural Energy*, Gateway Books, Bath, UK. Also Frokjaer-Jensen, B.(1981), "The Scandinavian

Research Organization and the Implosion Theory (Viktor Schauberger)," *Proc. First International Symposium on Nonconventional Energy Technology*, Toronto, pages 78-96.

Alexeff, I. and M. Radar (1995), "Possible Precursors of Ball Lightning - Observation of Closed Loops in High-Voltage Discharges," *Fusion Tech.* 27, pages 271-273.

Closed current loops were photographed during high voltage discharges. The loops enclose a magnetic field of very high energy density. They contract and quickly become compact force-free loops that superficially resemble spheres. In these toroidal geometries, the trapped internal magnetic field balances the external magnetic field to provide an almost force-free configuration. The bibliography cites numerous references on ball lightning.

Aspden, H.(1990), "Switched Reluctance Motot with Full Accommutation," U.S. Patent No. 4,975,608; ... (1993), "The World's Energy Future," *Proc. Int. Sym. on New Energy*, pages 1-19.

Barak, M.(1980), *Electrochemical Power Sources*, I.E.E. and Peter Peregrinus LTD., NY, pages 188-190.

Barber, B.P. and S.J. Putterman (1991), "Observation of synchronous picosecond sonoluminescence," *Nature* 353, pages 318-320; ... (1992), "Light Scattering Measurements of the Repetitive Supersonic Implosion of a Sonoluminescing Bubble," *Phys. Rev. Lett.* 69, pages 3839-42.

Barbour, J.(2000), *The End of Time*, Oxford University Press, NY.

A physicist proposal that the next revolution is physics will recognize the passage of time does not physically exist but is really an illusion of consciousness. Barbour shows how a timeless "best fit" hyper-surface algorithm is mathematically equivalent to the standard Hamiltonian time-evolving dynamics for all descriptions of physics including Newtonian mechanics, general relativity and quantum mechanics. It yields a many-worlds model without the passage of time.

Bearden, T.E.(1984), "Tesla's Electromagnetics and Its Soviet Weaponization," *Proc. Tesla Centennial Sym.*, International Tesla Society, Colorado Springs, pp 119-138.

Describes "scalar wave" technology. A scalar wave is defined as a propagating organization in the ZPE created by abruptly bucking electromagnetic fields.

Bearden, T.E. and T. Herold, E. Mueller (1985), "Gravity Field Generator Manufactured by John Bedini," Tesla Book Co., Greenville TX.

Experiments showed that Bedini's modified Kromrey, "G-Field" device produced "cold current."

Bearden, T.E.(1986), *Fer-De-Lance: A Briefing on Soviet Scalar Electromagnetic Weapons*, Tesla Book Co., Millbrae, CA., pages 107-108.

Bearden, T.E.(2002), *Energy from the Vacuum*, Cheniere Press, Santa Barbara, CA.
Epic work of over 950 pages details concepts and principles for extracting zero-point energy for electrical power generation.

P. Beckmann (1990), "Electron Clusters," *Galilean Electrodynamics*, vol 1, no 5, pp 55-58.
Explains how Shoulders' EV can be stabilized by a polarization interaction with the adjacent dielectric.

Bedini, J.(1991), "The Bedini Free Energy Generator," *Proc. 26th IECEC* vol 4, pages 451-456.

Bockris, J.(1996), "The Complex Conditions Needed to Obtain Nuclear Heat from D-Pd Systems," *J. New Energy* 1(3), pages 210-218.
The hypothesis is proposed that internal cracking of the cathode palladium (or nickel) is the needed triggering mechanism to obtain cold fusion or transmutation events. It explains why, even though it takes only three hours to load palladium rods to saturation, there can be delays of hundreds of hours before heat bursts occur. If the cracks should reach the surface, the deuterium fugacity is diminished and the reaction stops. Thin palladium nickel alloys or layers, as in Patterson's beads, allow the internal cracking to occur quickly giving reliable and repeatable results.

Bostick, W.H.(1957), "Experimental Study of Plasmoids," *Phys. Rev.* 106(3), page 404; ... (October 1957), "Plasmoids," *Scientific American* 197, page 87.

Bostick, W.H.(1966), "Pair Production of Plasma Vortices," *Phys. Fluids* 9, pages 2078-80.

Boyer, T.H.(1975), "Random Electrodynamics: The theory of classical electrodynamics with classical electromagnetic zero-point radiation," *Phys. Rev. D* 11(4), pages 790-808; ... (1969), "Derivation of Blackbody Radiation Spectrum without Quantum Assumptions," *Phys. Rev.* 182(5), pages 1374-83.

Boyer, T.H. (1976), "Equilibrium of random classical electromagnetic radiation in the presence of a nonrelativistic nonlinear electric dipole oscillator," *Phys. Rev. D* 13(10), pages 2832-45.

Brown, P.M.(1989), "Apparatus for Direct Conversion of Radioactive Decay Energy to Electrical Energy," U.S. Patent No. 4,835,433; ... (1987), "The Moray Device and the Hubbard Coil were Nuclear Batteries," *Magnets* 2(3), pages 6-12; ... (1990), "Tesla Technology and Radioisotopic Energy Generation," *Proc. Int. Tesla Sym.*, Colorado Springs, chap 2, pp 85-92; ...(1994), *Isotopic Power, a Summary of Nuclear Battery Development*, privately published.

Brown created a five watt nuclear battery using a weak (one Curie) radioactive source, Krypton 85. Since the radioactive source could only provide at best five milliwatts, Brown created an anomalous self running energy device. Brown used an LC oscillator, where the radioactive material ionized a corona around the coil. If the circuit is tuned to the ion-acoustic resonance of the corona, then the ion-oscillations could couple a ZPE coherence directly to the circuit.

Brown, P.M.(1997), *The Moray Radiant Energy Device: Operational Parameters, Design Criteria, and Considerations*, Aztec Publishing, Nashville, TN.

Summary of T.H. Moray's research includes an analysis of the tubes in Moray's patent (U.S. 2,460,707) as well as proposed simplified equivalent circuits.

Burridge, G.(1979), "The Smith Coil," *Psychic Observer* 35(5), pages 410-416.

Bush, R.T. (1992), "A Light Water Excess Heat Reaction Suggests that 'Cold Fusion' is 'Alkali-Hydrogen Fusion'," *Fusion Tech.* 22, page 287.

Celenza, L.S. and V.K. Mishra, C.M. Shakin, K.F. Liu (1986), "Exotic States in QED," *Phys. Rev. Lett.* 57(1), page 55; Caldi, D.G. and A. Chodos (1987), "Narrow e^+e^- peaks in heavy-ion collisions and a possible new phase of QED," *Phys. Rev. D* 36(9), page 2876; Jack Ng, Y. and Y. Kikuchi (1987), "Narrow e^+e^- peaks in heavy-ion collisions as possible evidence of a confining phase of QED," *Phys. Rev. D* 36(9), page 2880; Celenza,L.S. and C.R. Ji, C.M. Shakin (1987), "Nontopological solitons in strongly coupled QED," *Phys. Rev. D* 36(7), pages 2144-48.

Childress, D.H.(1991), *Vimana Aircraft of Ancient India and Atlantis*, Adventures Unlimited Press, Kempton, IL.

Cieplak, M. and L.A. Turski (1980), "Magnetic solitons and elastic kink-like excitations in compressible Heisenberg chain," *J. Phys. C: Solid State Physics* 13, pages L 777-780.

Cole, D.C. and H.E. Puthoff (1993), "Extracting energy and heat from the vacuum," *Phys. Rev. E* 48(2), pages 1562-65.

Theoretical proof that tapping the vacuum energy is possible and permissible within the laws of thermodynamics.

Cooke, M.B.(1983), *Einstein Doesn't Work Here Anymore*, Marcus Books, Toronto.

Presents a hyperspace theory where the noble gas nuclei constitute interdimentional node points for channeling energy.

Correa, P.N. and A.N. Correa (1995), "Electromechanical Transduction of Plasma Pulses," U.S. Patent 5,416,391. "Energy Conversion System," U.S. Patent 5,449,989; ... (1996), "XS NRG™ Technology," *Infinite Energy*, vol 2, no 7, pages 18-38; ... (1997), "Metallographic & Excess Energy Density Studies of LGEN™ Cathodes Subject to a PAGD Regime in Vacuum," *Infinite Energy*, vol 3, no 17, pages 73-78.

Fundamental discovery that a plasma tube tuned to operate at the abnormal glow discharge region exhibits an over unity energy gain. The abnormal glow discharge is a glow plasma that surrounds the cathode just before a vacuum arc discharge (spark) that occurs when the tube is slowly charged with increasing voltage. The abnormal glow exhibits a negative resistance characteristic as the tube begins an arc discharge. The patents illustrate how to make an appropriate charging circuit to cycle the tube in the abnormal glow discharge regime, control the cycle frequency, avoid the losses of the vacuum arc discharge, and rectify the excess energy onto batteries.

Crum, L.A.(1995), "Bubbles Hotter than the Sun," *New Scientist*, pages 36-40. Summarized in *Fusion Facts* 6(12), June 1995, page 10.

Overviews recent research in sonoluminescence including Hiller's discovery that the presence of noble gases make the luminosity increase by a factor of 30.

Davidson, D.A.(1990), *Energy: Breakthroughs to New Free Energy Devices*, Rivas, Greenville, TX, pages 16-18.

Davidson, J.(1989), *The Secret of the Creative Vacuum*, C.W. Daniel Co. Ltd., Essex, UK, pages 258-262.

Davidson, R.C.(1990), *Physics of Nonneutral Plasmas*, Addison-Wesley, NY.

Complete introduction to the topic of nonneutral (or charged) plasmas with abundant references to the scientific literature. The text provides the foundation for the mathematical modeling of nonneutral plasmas.

Davies, P.(1995), *About Time*, Simon & Schuster, NY.

Overview on the strange nature of time from the perspective of general relativity and quantum mechanics. Includes description of experiments where photons exhibit non-local connections across time as well as space.

DePalma, B.E. and C.E. Edwards (1973), "The Force Machine Experiments," privately published.

Deutsch, D.(1997), *The Fabric of Reality*, Penguin Books, NY.

The leading proponent of the many-worlds interpretation discusses quantum mechanics, epistemology, theory of computation, and the theory of evolution yielding a view of reality in which the past, present and future simultaneously exist.

Dirac, P.A.(1930), *Roy. Soc. Proc.*126, page 360. Also Gamow, G.(1966), *Thirty Years that Shook Physics*, Double Day, NY.

Dirac's virtual charge theory of the vacuum.

Ditmire, T.(1997), et al., "High Energy Explosion of Atomic Clusters: Transition from Molecular to Plasma Behavior," *Phys. Rev. Lett.*, vol 78, no 14, pp 2732-35.

Femtosecond laser pulse excitation of atomic noble gas clusters produce ion kinetic energies three orders of magnitude higher than expected.

Dollard, E.(1988), "Van Tassel's Caduceus Coils," private communication.

Van Tassel experimented with numerous caduceus coils that often contained quartz cores. His notes stated that the cross-over angle for the two opposing windings should be 22.5 degrees.

Dufour, J.(1993), "Cold Fusion by Sparking in Hydrogen Isotopes," *Fusion Technology* 24, pages 205-228.

Dwivedi, B.N. and K.J.H. Phillips (June 2001), "The Paradox of the Sun's Hot Corona," *Sci. Amr.* 84(6) pages 40-47.

Describes the sun's corona as a turbulent, fractal distribution of solar flares, which are plasma vortex filaments. When the opposing magnetic fields of the filaments cross and cancel each other, there can be a sudden release of up to 10^{25} joules. Could the anomalously high energy arise from the zero-point energy?

Eberlien, C. (1996), "Sonoluminescence as Quantum Vacuum Radiation, *Phys. Rev. Lett.*, vol 76, pp 3842-45; ...(1996), "Theory of quantum radiation observed as sonoluminescence,"

Phys. Rev. A, vol 53, pp 2772-87.

Einstein, A., B. Podolsky, N. Rosen (1935), "Can Quantum Mechanical Description of Physical Reality be Considered Complete?" *Phys. Rev.*, vol 47, p 777.

The original paper which showed that in quantum mechanics particles born from the same quantum event remain strongly correlated even when separated. Experiments in the 1980's confirmed it. The essence of the paradigm shift from classical physics is that the elementary particles cannot be "locally" modeled. See G. Zukav (1979), *The Dancing Wu Li Masters*, Bantam Books, NY, for a thorough discussion.

Egely, G.(1986), "Energy Transfer Problems of Ball Lightning," Central Research Institute for Physics, Budapest, Hungary.

Everett, H.(1957), "Relative State Formulation of Quantum Mechanics," *Rev. Mod. Phys.* 29 (3), page 457. Also 1973, B.S. Dewitt N. Graham, *The Many Worlds Interpretation of Quantum Mechanics,* Princeton University Press.

Everett was the first to introduce a self-consistent formulation of quantum mechanics without invoking postulates regarding the observer. The formulation yields a superspace containing an infinite number of three dimensional universes.

Greer, S.M.(2001), *Disclosure*, Crossing Point Inc., Crozet, VA;

Numerous testimonies from engineers and witnesses that zero-point energy devices including gravitational propulsion have been developed on top secret projects.

Farnsworth, P.T.(1939), "Cold Cathode Electron Discharge Tube," U.S. Patent 2,184,910.

Cup shaped electrodes provide a cold cathode discharge which has some similarity to the hollow cathode discharge.

Feynman, R.P.(1985), *QED The Strange Theory of Light and Matter,* Princeton Univ. Press, Princeton, NJ; ... (1949), "Space-Time Approach to Quantum Electrodynamics," *Phys. Rev.*, vol 76, p 769.

Forward, R.L.(1984), "Extracting electrical energy from the vacuum by cohesion of charged foliated conductors," *Phys. Rev. B* 30(4), pages 1700-2.

Fox, H. and R.W. Bass, S.X.Jin (1996), "Plasma-Injected Transmutation," *J. New Energy* 1(3), pages 222-230.

Classical calculation showing that nuclear charge clusters can be produced at low energy and yet gain sufficient acceleration for their contained protons to penetrate the Coulomb barrier and transmute lattice nuclei.

Friedman, J.L. and R.D. Sorkin (1982), "Half-Integral Spin from Quantum Gravity," *Gen. Rel. Grav.*, vol 14, no 7, pp 615-620.
Decribes the topology change required to define the spin ½ particles.

Friedman, N.(1990), *Bridging Science and Spirit*, Living Lake Books, St. Louis, MO.
Overview of the common elements in David Bohm's physics, Ken Wilber's Perennial Philosophy and Jane Robert's Seth material yields a unified view of physical reality that includes nonlocal connectivity, consciousness, and higher dimensional space.

Gluck, P.(1992), Letters to the editor, *Fusion Facts* 4(7), pages 22-24.

Graneau, P. and P.N. Graneau (1985), "Electrodynamic Explosions in Liquids," *Appl. Phys. Lett.* 46(5), pages 468-470.

Graneau, P. (1997), "Extracting Intermolecular Bond Energy from Water," *Proc. Fourth Int. Sym. on New Energy*, pp 65-70; also *Infinite Energy*, vol 3, no 3-4, pp 92-95.
High speed photography reveals a plasma ball sitting in the accelerator muzzle of the water arc explosion experiments.

Gray, E.V.(1976), "Pulsed Capacitor Discharge Electric Engine," U.S. Patent No. 3,890,548.

Gray, E.V.(1987), "Efficient Electrical Conversion Switching Tube Suitable for Inductive Loads," U.S. Patent 4,661,747; ...(1986), "Efficient Power Supply Suitable for Inductive Loads," U.S. Patent 4,595,975.

Gribbin, J.(1995), *Schrodinger's Kittens and the Search for Reality*, Back Bay Books, NY.
Overview of the foundational issues (non-local connectivity) and interpretations of quantum mechanics including the Copenhagen interpretation, many-worlds interpretation, and the transactional interpretation where distant quantum entangled objects can be instantaneously connected across both space and time.

Gundersen, M.A. and G. Schaefer (1990), *Physics and Applications of Pseudosparks*,

Plenum Press, NY.

Conference proceedings studying hollow cathode discharge phenemena. If a glow discharge is created in the interior of a hollow cathode, tremendous currents may be switched and triggered by a small external signal. The phenomena generates an intense beam of ions between the electrodes. The initial stage of the discharge exhibits anomalous behavior including a negative resistance characteristic that is a powerful version of the abnormal glow discharge.

Hadley, M.J. (1997), "The Logic of Quantum Mechanics Derived From Classical General Relativity," *Found. Phys. Lett.*, vol 10, no 1, pp 43-60; ... (1996), "A Gravitational Explanation of Quantum Mechanics,"

A pure general relativity solution involving closed time-likes curves is used to model the electron as a four dimensional "geon," which fulfills the non-distributive logic of quantum mechanics. The solution involves a self-interaction across the time dimension (essentially via wormholes) that results in the probabilistic outcomes characteristic of quantum mechanics.

Haisch, B. and A. Rueda, H.E. Puthoff (1994), "Inertia as a zero-point field Lorentz force," *Phys. Rev. A* 49(2), pages 678-694.

Hathaway, G.(1991), "Zero-Point Energy: A New Prime Mover? Engineering Requirements for Energy Production & Propulsion from Vacuum Fluctuations," *Proc. 26th IECEC* vol 4, pages 376-381.

Hiller, R. and S. Putterman, B. Barber (1992), *Phys. Rev. Lett.* 69, pages 1182-84.

Honig, W.M.(1986), *The Quantum and Beyond*, Philosophical Library, NY; ... (1974), "A Minimum Photon Rest Mass using Planck's Constant and Discontinuous Electromagnetic Waves," *Found. Phys.* 4(3), pages 367-380.

Hubbard, A.M.(1919), Web site:

Alfred Hubbard at age 19 appears to be the first inventor in history to demonstrate a self-running, electrical "free energy" device. The device appears to have been replicated by Paul Brown (1989), who apparently disguised it as a nuclear battery.

Huntley, H.E. (1970), *The Divine Proportion*, Dover Publications, NY.

Hyde, W.W.(1990), "Electrostatic Energy Field Power Generating System," U.S. Patent No. 4,897,592. The invention is summarized in King (1991).

Jandel, M.(1990), "Cold Fusion in a Confining Phase of Quantum Electrodynamics," *Fusion Tech.* 17, pages 493-499.

Jennison, R.C.(1978), "Relativistic Phase-Locked Cavities as Particle Models," *J. Phys. A Math Gen. VII*(8), pages 1525-33; ... (1989), "A New Classical Relativistic Model of the Electron," *Phys. Lett. A* 141(8/9), pages 377-382.

Jennison,R.C.(1990), "Relativistic Phase-Locked Cavity Model of Ball Lightning," Electronics Laboratory, University of Kent, U.K.

Jin, S.X. and H. Fox (1997), "Characteristics of High-Density Charge Clusters: A Theoretical Model," *J. New Energy*, vol 1, no 4, pp 5-20.

A mathematical model of charged clusters (Shoulder's EV's) is presented that shows the stability is due to a helical vortex ring possessing an extraordinary poloidal circulation. In this nonrelativistic calculation, the poloidal filament would have to be thin. A spherical electron cluster is unstable and would tend to form into a toroid by a force balance relationship. The calculation shows that the energy density of a charge cluster is a hundred times higher than in a supernova explosion.

Johnson, G.L.(1992), "Electrically Induced Explosions in Water," *Proc. 27th IECEC* vol 4, pages 4.335-338.

Johnson, P.O.(1965), "Ball Lightning and Self Containing Electromagnetic Fields,"*Am. J. Phys.* 33, page 119.

Kaku, M. (1994), *Hyperspace*, Anchor Books Doubleday, NY.

Layman's overview to the hyperspace theories of theoretical physics includes general relativity, nuclear standard model, quantum gravity, superstrings, many-worlds, wormholes, time warps, Hawking's universal wave function and Coleman's parallel universes. Shows that unification of physics elegantly occurs via the hyperspace theories. It is notewothy that the fundamental action in the unifying theories is occurring at the Planck length (10^{-33} cm) setting the stage for a microscopic theory of the vacuum fluctuations.

Kalinin, Yu G. et al.(1970), "Observation of Plasma Noise During Turbulent Heating," *Sov. Phys. Dokl.* 14(11), page 1074; Iguchi, H.(1978), "Initial State of Turbulent Heating of Plasmas,"_*J. Phys. Soc. Jpn.* 45(4), page 1364; Hirose, A.(1974), "Fluctuation Measurements in a Toroidal Turbulent Heating Device," *Phys. Can.* 29(24), page 14.

Kiehn R.M. (1977), "Periods on Manifolds, Quantization and Gauge," *J. Math Phys.* 18(4); also

The notions of quantized flux, charge, and spin can be derived from topological ideas of fields built on manifolds. A non Euclidean topology is required for the quantization where the values of the closed integrals on a closed oriented manifold are integer multiples of some smallest value. The flux quantum, charge quantum and angular momentum quantum are not independent but are related by a topological constraint.

Kiehn R.M. (1998), "Electromagnetic Wave in the Vacuum with Torsion and Spin,"

Torsion and spin wave solutions to Maxwell's equations are derived that are gauge invariant. Values of the spin integral form rational ratios giving a classical solution supporting quantized angular momentum having qualities like the photon.

Kiehn R.M. (1997), "Continuous Topological Evolution,"

Cartan calculus is used to model continuously changing topology applicable to reversible and irreversible phenomena. It models hole and handle creation/annihilation, turbulence, chaos, action viewed abstractly in terms of kinetic fluctuations, thermodynamics, topological vorticity and parity, harmonic forms and their relationship to angular momentum, probability current, evolution of defects, links and knots, and quantization. The production of torsion defects is the key to the understanding of large scale structures in continuous media. The creation of topological torsion involves discontinuous processes or shocks. The Cartan method is explicitly coordinate free, metric free, and connection free.

King, M.B.(1984), "Macroscopic Vacuum Polarization," *Proc. Tesla Centennial Symposium*, International Tesla Society, Colorado Springs, pages 99-107. Also (1989), pages 57-75.

King, M.B.(1986), "Cohering the Zero-Point Energy," *Proc. of the 1986 International Tesla Symposium*, Colorado Springs, section 4, pages 13-32. Also (1989), pages 77-106.

King, M.B.(1989), *Tapping the Zero-Point Energy*, Paraclete Publishing, Provo, UT; also Advenutures Unlimited Press, Kempton, IL; ... (1991), "Tapping the Zero-Point Energy as an Energy Source," *Proc. 26th IECEC* vol 4, pages 364-369; ... (1993), "Fundamentals of a Zero-Point Energy Technology," *Proc. Int. Sym. on New Energy*, pages 201-217.

King, M.B.(1994), "Vacuum Energy Vortices," *Proc. Int. Sym. on New Energy*, pages

257-269.
Discusses force-free vortices. Suggests that counter-rotating plasmas can induce a ZPE coherence akin to a large scale "pair production."

King, M.B.(1996), "The Super Tube," *Proc. Int. Sym. on New Energy*, pages 209-221; also *Infinite Energy* 2(8), pages 23-28.
A powerful plasma tube intended to cohere the ZPE combines the use of a hollow cathode discharge, radioactive cathodes and inert gas mixtures; it operates in the abnormal glow discharge regime. Output energy in the form of large voltage spikes are efficiently absorbed by a pulse current multiplier (PCM) circuit which might offer a solid state means of tapping the vacuum energy.

King, M.B.(1997), "Charge Clusters: The Basis of Zero-Point Energy Inventions," *J. New Energy*, vol 2, no 2, pp 18-31; also *Infinite Energy*, vol 3, no 3-4, pp 96-102.
Shows how many "free energy" inventions utilize (sometimes unwittingly) the phenomena of charge clusters, which may provide the coupling to the zero-point energy for their source of power.

King, M.B.(1999), "Vortex Filaments, Torsion Fields and the Zero-Point Energy," *J. New Energy*, vol 3, no 2/3, pp 106-116.
Overview of Russian research on torsion fields suggest plasma vortex action can induce large vacuum torsion fields in the ZPE.

King, M.B. (2001), *Quest for Zero-Point Energy*, Adventures Unlimited Press, Kempton, IL.
Collection of author's technical papers from 1991 to 2001 describing engineering principles for tapping the vacuum energy.

Kiwamoto, Y. and H. Kuwahara, H. Tanaca (1979), "Anomalous Resistivity of a Turbulent Plasma in a Strong Electric Field," *J. Plasma Phys.* 21(3), page 475.

Kozyrev, N.A.(1968), "Possibility of Experimental Study of the Properties of Time," Joint Publication Research Service, Arlington VA.

Kuhn, T.S.(1970), *The Structure of Scientific Revolutions*, University of Chicago Press, Chicago.
Shows that throughout history scientific paradigm shifts have been strongly resisted.

Lagarkov, A.N. and I.M. Rutkevich (1994), *Ionization Waves in the Electrical Breakdown of Gases*, Springer-Verlag, NY.

La Violette, P.A.(1985), "An introduction to subquantum kinetics...," *Intl. J. Gen. Sys.* 11, pages 281-345; ... (1991), "Subquantum Kinetics: Exploring the Crack in the First Law," *Proc. 26th IECEC* vol 4, pages 352-357.

Lerner, E.(2002), Focus Fusion Society,

New version of the zeta pinch device creates a plasmoid exhibiting temperatures over one billion degrees K. It can trigger hydrogen boron fusion with no radioactive by-products and output power directly as electricity. Could the plasmoid also induce a zero-point energy coherence?

Lewis, E.H.(1995), "Tornados and Ball Lightning," *Extraordinary Science* VII (4), pages 33-37; ...(March 1996), "Tornados and Tiny Plasmoid Phenomena," *New Energy News* 3(9), pages 18-20; ...(Feb 1994), "Some Important Kinds of Plasmoid Traces Produced by Cold Fusion Apparatus," *Fusion Facts* 6(8), pages 16-17.

Overview of plasmoid phenomena including tornados, ball lightning, and microscopic EV's. Includes an abundant list of references.

Lindemann, P.(2001), *The Free Energy Secrets of Cold Electricity*, Clear Tech Inc., Metaline Falls, WA;

Proposes that an abrupt, unipolar, electrical discharge can launch a "cold" form of electricity that does not heat the conducting wires. The phenomena was first observed by Tesla and was utilized by Edwin Gray (1975) in the discharge tube driving his motor.

Mallove, E.F.(1992), "Protocols for Conducting Light Water Excess Energy Experiments," *Fusion Facts* 3(8), page 15; Noninski, V.C.(1992), "Excess Heat during the Electrolysis of a Light Water Solution of K_2CO_3 with a Nickel Cathode," *Fusion Tech.* 21, pages 163-167.

Matthey, P.H. (1985), "The Swiss ML Converter - A Masterpiece of Craftsmanship and Electronic Engineering," in H.A. Nieper (ed.), *Revolution in Technology, Medicine and Society*, MIT Verlag, Odenburg.

McElrath, H.B.(1936), "Electron Tube," U.S. Patent 2,032,545.

Patents the use of radioactive materials to create a cold cathode.

Medvedeff, N.J.(1961), *Nuclear Dynamics*, privately published, Hanover, MA.

Meyer, S.L.(1991), *The Birth of a New Technology*, Water Fuel Cell, Grove City, OH; ... (1989), "Controlled Process for the Production of Thermal Energy from Gases and Apparatus Useful Therefore," U.S. Patent No. 4,826,581; ... (1990), "Method for the Production of a Fuel Gas, (Electrical Polarization Process)," U.S. Patent No. 4,936,961.

Mesyats, G.A.(1996), "Ecton Processes at the Cathode in a Vacuum Discharge," *Proc. 17th International Symposium on Discharges and Electrical Insulation in Vacuum*, pages 720-731.

Russian research is presented analyzing their discovery of charge clusters, called "ectons." Ectons often arise from micro explosions on the surface of the cathode, where surface imperfections such as microprotrusions, adsorbed gases, dielectric films and inclusions play an important role. The simplest way to initiate ectons is to cause an explosion of cathode microprotrusions under the action of field emission current. Experiments confirm microprotrusions jets can form from liquid or melting metal. The breakdown of thick dielectric films in their charging with ions also plays an important role in the initiation of ectons. A commonly used way to initiate an ecton is to induce a vacuum discharge over a dielectric in contact with a pointed, metal cathode. An ecton can readily be excited at a contaminated cathode with a low density plasma, but a clean cathode requires a high plasma density.

Michrowski, A.(1993), "Vacuum Energy Developments," *Proc. Int. Sym. on New Energy*, pages 407-417.

Mills, R.L. and S.P. Kneizys (1991), "Excess Heat Production by the Electrolysis of an Aqueous Potassium Carbonate Electrolyte and the Implications for Cold Fusion," *Fusion Tech.* 20, pages 65-81.

Milonni, P.W. and R.J. Cook, M.E. Goggin (1988), "Radiation pressure from the vacuum: Physical interpretation of the Casimir force," *Phys. Rev. A* 38(3), pages 1621-23.

Mizuno, T. and M. Enyo, T. Akimoto, K. Azumi (1993), "Anomalous Heat Evolution from $SrCeO_3$ - Type Proton Conductors during Absorption/Desorption of Deuterium in Alternate Electric Field," *4th Int. Conf. on Cold Fusion.* Abstract in *Fusion Facts*, Dec. 1993, page 30.

Misner, C. and K. Thorne, J. Wheeler (1970), *Gravitation*, W.H. Freeman, NY.

Thorough text on general relativity and differential forms. Chapter 43 discusses the zero-point energy and geometrodynamics. Chapter 44 discusses the problems of modeling actual charge and physics beyond the singularity of gravitational collapse.

Moore, C.B.(1971), *Keely and His Discoveries*, Health Research, Mokelumne Hill, CA.

Moray, T.H.,(1949) "Electrotherapeutic Apparatus," U.S. Patent 2,460,707.
Moray's only patent. It contains three tubes that do not fit the apparatus. These seem to match the oscillator and valve tubes appropriate for his radiant energy device.

Moray, T.H. and J.E. Moray (1978), *The Sea of Energy*, Cosray Research Institute, Salt Lake City, UT.
The story and description of T. Henry Moray's radiant energy invention. It includes numerous letters of testimony from witnesses.

Moreland, J.W.(1997), "An Update on the Continuing Research into T.H. Moray and Other Free Energy Devices with Conclusions," Aztec Publishing, Bethpage, RI;
Analysis of Moray's research suggests that radioactive fission is the primary source of energy.

Ouspensky, P.D.(1970), *Tertium Organum*, Vintage Books, NY.
Philosophical treatise written in 1912 which proposes that the perception of time is a result of consciousness partial perception of a higher physical dimension. For human consciousness, a spatial fourth dimension exists, and our partial perception of it is experienced as the passage of time. Since one extra physical dimension is sufficient to contain an infinite number of three dimensional universes, this model sets the foundation for a timeless many-worlds interpretation of reality. It appears to be a model to which many of today's quantum physicists/philosophers are heading, and it is remarkable that it was publish twenty years before the discovery of quantum mechanics.

Panici, D.(1992), private communication.
Research into energetic anomalies from pulse charging lead acid batteries. A fresh battery (that has never been discharged) is required to yield a "cold current" observation.

Panov V.F., et al.(1998), "Torsion Fields and Experiments," *J. New Energy* 2(3-4), pages 29-39.
Overviews the history and development of the torsion field research. Summarizes the experimental program at Perm University which includes the effect of torsion fields on chemical and electrochemical processes, crystal growth, nuclear transmutation, biological systems, medical applications, and

extraction of energy from the vacuum.

Papp, J.(1984), "Inert Gas Fuel, Fuel Preparation Apparatus and System for Extracting Useful Work from the Fuel," U.S. Patent No. 4,428,193; ...(1972), "Method and Means of Converting Atomic Energy into Utilizable Kinetic Energy," U.S. Patent No. 3,670,494. Also (2003), *Infinite Energy*, vol 9, issue 51, pages 6-63.

Noble gas engine drives its piston from plasma discharge, which manifests anomalously excessive energy.

Pappas, P.T.(1991), "Energy Creation in Electrical Sparks and Discharges: Theory and Direct Experimental Evidence," *Proc. 26th IECEC* vol 4, pages 416-423.

Patterson, J.(1996), U.S. Patent 5,494,559.

The first U.S. "cold fusion" patent with the claim of excess energy (2000%) granted. The Patterson cell contains hundreds of sub millimeter beads made in a electroplating process of depositing multiple, alternating, thin layers of very pure nickel and palladium. It runs using a light water electolyte. It is considered the best of the electrolytic cold fusion type of experiments with complete repeatability and a record gain of over 1000 in one demonstration.

Peat, E.D.(1988), *Superstrings and the Search for the Theory of Everything*, Contemporary Books, Chicago.

Perreault, B.A.(1999), *Radiant Energy Power Generation*, Nu Energy Horizons, Rumney, NH;

Analysis of T. Henry Moray's energy device suggests that fission was the dominant source of energy. Details the author's own experiments.

Piantelli, F.(1995), "Energy generation and generator by means of anharmonic stimulated fusion," International Patent WO 95/20816. Summarized by W. Collis, *Fusion Facts* 7(6), Dec 1995, pages 14-15.

Light water cold fusion experiment using pure crystalline nickel that is fully loaded. Once the cell is triggered by a vibrational stress, ultrasonic stationary waves maintain the reaction. The reaction will cease if the temperature rises sufficiently to destroy the crystalline structure of the core.

Preparata, G.(1991), "A New Look at Solid-State Fractures, Particle Emission and Cold Nuclear Fusion," *Il Nuovo Cimento* 104 A(8), page 1259; ... (1990), "Quantum field theory of superradiance," in Cherubini, R., P. Dal Piaz, B. Minetti (editors), *Problems of Fundamental Modern Physics*, World Scientific, Singapore.

Prigogine, I. and G. Nicolis (1977), *Self Organization in Nonequilibrium Systems,* Wiley, NY; Prigogine, I. and I. Stengers (1984), *Order Out of Chaos*, Bantam Books, NY.

Puharich, A.(1981), "Water Decomposition by Means of Alternating Current Electrolysis," *Proc. First International Symposium on Nonconventional Energy Technology,* Toronto, pages 49-77; ... (1983), "Method and Apparatus for Splitting Water Molecules," U.S. Patent No. 4,394,230.

Puthoff, H.E.(1987), "Ground state of hydrogen as a zero-point fluctuation determined state," *Phys. Rev. D* 35(10), pages 3266-69.

Puthoff, H.E.(1989), "Gravity as a zero-point fluctuation force," *Phys. Rev. A* 39(5), pages 2333-42.

Puthoff, H.E.(1990), "The energetic vacuum: implications for energy research," *Spec. Sci. Tech.* 13(4), pages 247-257.

Ralphs, J.D.(1993), *Exploring the Fourth Dimension*, Lewellyn, St. Paul, MN.

Describes how a Euclidian fourth dimension resolves many of the paradoxes in physics as well as begin to explain paranormal phenomena. Uses perception theory from psychology to show that how higher dimensions could physically exist without our awareness of them.

Rausher, E.A.(1968), "Electron Interactions and Quantum Plasma Physics," *J. Plasma Phys.* 2(4), page 517.

Perhaps the first theoretical paper to suggest zero-point energy coherence can occur in plasma.

Reed, D. (1992), "Toward a Structural Model for the Fundamental Electrodynamic Fields of Nature," *Extraordinary Science IV*(2), pages 22-33; ... (1993), "Evidence for the Screw Electromagnetic Field in Macro and Microscopic Reality," *Proc. Int. Sym. on New Energy* , pages 497-510; ... (1994), "Beltrami Topology as Archetypal Vortex," *Proc. Int. Sym. on New Energy,* pages 585-608; ...(1996), "The Beltrami Vector Field - The Key to Unlocking the Secrets of Vacuum Energy?" *Proc. Int. Sym. on New Energy*, pages 345-363; ... (1997), "The Vortex as Topological Archetype - A Key to New Paradigms in Physics and Energy Science," *Proc. Fourth Int. Sym. on New Energy*, pp 207-224; ... (1998), "Torsion Field Research," *New Energy News*, vol 6, no 2, p 22; ... (1998), "Excitation and Extraction of Vacuum Energy Via EM - Torsion Field Coupling - Theoretical Model," *J. New Energy*, (to be published).

Reinhardt, J. and B. Muller, W. Greiner (1980), "Quantum Electrodynamics of Strong Fields in Heavy Ion Collisions," *Prog. Part. and Nucl. Phys.* 4, page 503.

Resines, J.(1989), "The Complex Secret of Dr. T. Henry Moray," Borderland Sciences, Gaberville, CA.

Roberts, J.(1977), *The Unknown Reality: A Seth Book, Volume 1 & 2*, Prentice-Hall, Englewood Cliffs, NJ.

Spiritually channeled dictation which presents a model of reality similar to Everett's many-worlds interpretation, Ouspensky's *Tertium Oraganum*, and Barbour's *The End of Time*. The model suggests time is an illusion of consciousness, and there exists an infinite number of probabilistic universes available for us to experience.

Roden, M.(1992), *Aero Dynamics*, Aero Dynamics Master Society, San Ysidro, CA.

B. Rubick (1999), "The Perennial Challenge of Anomalies at the Frontiers of Science," *Infinite Energy*, vol 5, issue 26, pp.34-41. Essay shows that paradigm violating experiments and anomalies are ferociously resisted by the scientific community in both the past and the present.

Rodrigues, W.A. and J.Y. Lu (1997), "On the Existence of Undistorted Progressive Waves (UPWs) of Arbitrary Speeds (Zero to Infinity) in Nature," *Found. Phys.*, vol 27, no 3, pp 435-508.

Samokhin, A.(1990), "Vacuum energy - a breakthrough?" *Spec. Sci. Tech.* 13(4), page 273.

Scheck, F.(1983), *Leptons, Hadrons and Nuclei*, North Holland Physics Publ., NY, pages 213-223.

Schwinger, J.(1993), "Casimir light: The source," *Proc. Natl. Acad. Sci. USA* 90, pages 2105-6.

Scully, M.O. and M. Sargent (March 1972), "The Concept of the Photon," *Physics Today*, page 38.

Sego, R.(1981), "The Moray Energy Device - Its Workings," privately published.

Analysis of T.H. Moray's device includes study of text books Moray referenced: G. Le Bon, *Evolution of Matter*, and E. Rutherford, *Radio-Activity*. Rutherford noted that a combination of radium, thorium and uranium would ionize twice as much gas as any one alone.

Seike, S. (1978), *The Principles of Ultrarelativity*, G-Research Laboratory, Tokyo, Japan.

Models a fourth dimensional energy flux that can be tapped by pulsed Mobius band coils.

Senitzky, I.R.(1973), "Radiation Reaction and Vacuum Field Effects in Heisenberg-Picture Quantum Electrodynamics," *Phys. Rev. Lett.* 31(15), page 955.

Sethian, J.D. and D.A. Hammer, C.B. Whaston (1978), "Anomalous Electron-Ion Energy Transfer in a Relativistic-Electron-Beam Heated Plasma," *Phys. Rev. Lett.* 40(7), page 451; Robertson, S. and A. Fisher, C.W. Roberson (1980), "Electron Beam Heating of a Mirror Confined Plasma," *Phys. Fluids* 32(2), page 318; Tanaka, M. and Y. Kawai (1979), "Electron Heating by Ion Acoustic Turbulence in Plasmas," *J. Phys. Soc. Jpn.* 47(1), page 294.

Shakhparonov, I.M.(1998), "Kozyrev-Dirac Emanation Methods of Detecting and Interaction with Matter," *J. New Energy* 2(3-4), pages 40-45.

Overviews concepts and experiments of creating "non-orientable" topological structures using conductive Mobius band circuit elements. Such topological structures have a hyperspatial nature and cannot be oriented in three dimensional space. The projection into three space manifest phenomena akin to ball lightning. Magnetic monopoles are likewise of this nature, and experiments are referenced where beams of such are launched, detected and shown to exhibit alterations in gravity as well as the pace of time.

Shoulders, K.R.(1991), "Energy Conversion Using High Charge Density," U.S. Patent No. 5,018,180.

Fundamental discovery on how to launch a micron size, negatively charged plasmoid called "Electrum Validum" (EV). An EV yields excess energy (over unity gain) whenever it hits the anode or travels down the axis of an hollow coil. The excess energy comes from the ZPE.

Shoulders, K. and Shoulders, S. (1996), "Observations on the Role of Charge Clusters in Nuclear Cluster Reactions," *J. New Energy* 1(3), pages 111-121.

Experimental evidence in the form of micrographs and X-ray microanalysis is presented suggesting that nuclear charge clusters, (micron sized plasmoids containing 10^{11} electrons and 10^{6} protons or deuterons) can accelerate into lattice nuclei with sufficient kinetic energy to overcome the Coulomb barrier and trigger transmutation events. The hypothesis to explain cold fusion is proposed where electrolytic loading of palladium or nickel causes cracking and fractoemission of the charge clusters.

Sigma, R. (1977), *Ether Technology*, Tesla Book Co., Millbrae CA.

Silberg, P.A.(1962), "Ball Lightning and Plasmoids," *J. Geophys. Res.* 67(12), page 4941.

Singer, S.(1971), *The Nature of Ball Lightning*, Plenum Press, NY;

Skvortsov, V. and N. Vogel, A. Lebedev (1996), "The Dynamic of Matter Transition into Extreme States Initiated by High Power Micro Beam of Heavy Ions," *Proc. 17th International Symposium on Discharges and Electrical Insulation in Vacuum*, pages 813-817.

Computer simulation of short pulse, intense micro beams of heavy ions interacting with condensed matter shows a shock wave formation of a tubular plug of dense matter at extemely high pressures that could induce fusion. The analysis supports Shoulder's hypothesis that cold fusion effects might be produce by accelerated ions carried by charge clusters.

Smith, T. (1997), "The Hyper Diamond Feynman Checkerboard,"

Theory based on Clifford algebra that yields the standard model, masses of the elementary particles, gravity, Bohm's implicate order and the many worlds interpretation of quantum mechanics.

Smith, W.B.(1964), *The New Science*, Fern-Graphic Publ., Mississauga, Ontario.

Smolin, L.(2001), *Three Roads to Quantum Gravity*, Basic Books, NY.

Overview of the theories of quantum gravity, which strive to unify general relativity with quantum electrodynamics. Space-time and the elementary particles are both derived from detailed modeling of the vacuum fluctuations.

Spence G.M.(1988), "Energy Conversion System," U.S. Patent No. 4,772,816.

Storms, E.(1991), "Review of Experimental Observations about the Cold Fusion Effect," *Fusion Tech.* 20, pages 433-477.

Suzuki, M.(1984), "Fluctuation and Formation of Macroscopic Order in Nonequilibrium Systems," *Prog. Theor. Phys. Suppl.* 79, pages 125-140; Hasegawa, A.(1985), "Self-Organization Processes in Continuous Media," *Adv. Phys. 34(1)*, pages 1-41; Firrao, S.(1984), "Physical Foundations of Self-Organizing Systems Theory," *Cybernetica*

17(2), pages 107-124; Haken, H.(1971), *Synergetics*, Springer Verlag, NY.

Sweet, F. and T.E. Bearden (1991), "Utilizing Scalar Electromagnetics to Tap Vacuum Energy," *Proc. 26th IECEC* vol 4, pages 370-375; ... (1988), "Nothing is Something: The Theory and Operation of a Phase-Conjugate Vacuum Triode;" ... (1989), private communication.

Valone, T. (1995), Electrogravitics Systems, Integrity Research Institute, Washington DC.

Collection of technical essays analyzing the research of T. Townsend Brown, who discovered that a highly charged capacitor exhibited a thrust that might be caused by artificial gravity.

Valone, T. (2002), *Harnessing the Wheelwork of Nature*, Adventures Unlimited Press, Kempton, IL.

Collection of technical essays from engineers and physicists analyzing the research of Nikola Tesla.

Vassilatos, G.(1991), *Lost Science*, Adventures Unlimited Press, Kempton, IL.

Collection of essays describing the research of historic inventors who made discoveries that violated the scientific paradigm. These include Reichenbach's "odic" energy, Meucci's telephony, Stubblefield's earth energy, Tesla's amplifying transmitter, Rife's microscope, Moray's radiant energy, Brown's electrogravitics, Graveau's infrasound, and Farnsworth's fusor reactor.

Walker, E.H.(2000), *The Physics of Consciousness*, Perseus Publishing, Cambridge MA.

Shows how the physiology of the brain can support a macroscopic quantum coherence, where spirit can control the wave function collapse. Includes a fine overview of the various interpretations of quantum mechanics.

Walton, A.J. and G.T. Reynolds (1984), "Sonoluminescence," *Adv. Phys.* 33(6), pages 595-600.

Wheeler, J.A.(1962), *Geometrodynamics*, Academic Press, NY.

J.A. Wheeler (1980), "Beyond the Black Hole," in H. Woolf (ed.), *Some Strangeness in the Proportion,* Addison-Wesley, Reading, MA, pp 341-375.

Discusses the strange wave-particle descriptions in quantum mechanics. A variation of the two slit diffraction experiments (known as "delayed choice") alter the experimental setup after the particle is launched. Quantum events seem to reach across time.

Winter, D.(1991), *The Star Mother, Geometric Keys to the Resonant Spirit of Biology*, Crystal Hill Farm Publications, Eden, NY.

Winterberg, F.(1990), "Maxwell's Equations and Einstein-Gravity in the Planck Aether Model of a Unified Field Theory," *Z. Naturforsch.* 45 a, pages 1102-16; ... (1991), "Substratum Interpretation of the Quark-Lepton Symmetries in the Planck Aether Model of a Unified Field Theory," *Z. Naturforsch.* 46 a, pages 551-559.

Wolf, F.A.(1990), *Parallel Universes*, Simon & Shuster, NY.

Woodhouse, M.B. (1996), *Paradigm Wars*, Frog Ltd., Berkeley CA.

Discussion of world views and paradigm shifts across many fields including system holism, system self-organization, new physics, non-locality, higher dimensions, energy, time, consciousness, transpersonal psychology, religion, Perennial philosophy, mysticism, health, education, environment and extraterrestrial contact.

Ziolkowski, R.W. and M.K. Tippett (1991), "Collective effect in an electron plasma system catalyzed by a localized electromagnetic wave," *Phys. Rev. A* 43(6), pages 3066-72.

Mathematical analysis of Shoulder's EV that includes a significant (vacuum polarization) displacement current term since the EV formation time is of the same order as the plasma frequency period. The resulting nonlinear Klein-Gordon equation contains vorticity terms and a term similar to a quantum mechanical potential which compensates for the repulsion. The system is solved by numerical methods for a stable, localized wave solution which matches the EV in size and charge density.

TAPPING THE ZERO-POINT ENERGY

How Free Energy & Anti-Gravity Might Be Possible With Today's Physics

by Moray B. King

King explains how free energy and anti-gravity are possible. The theories of the zero point energy maintain there are tremendous fluctuations of electrical field energy embedded within the fabric of space: How in the 1930s inventor T. Henry Moray could produce a fifty kilowatt "free energy" machine; how an electrified plasma vortex creates anti-gravity; how the Pons/Fleischmann cold fusion experiment could produce tremendous heat without fusion; and how certain experiments might produce a gravitational anomaly; Tapping the Zero-Point Energy as an Energy Source; Fundamentals of a Zero-Point Energy; Technology Vacuum Energy Vortices; The Super Tube Charge Clusters: The Basis of Zero-Point Energy Inventions; Vortex Filaments, Torsion Fields and the Zero-Point Energy; Transforming the Planet with a Zero-Point Energy; Experiment Dual Vortex Forms: The Key to a Large Zero-Point Energy Coherence; plus tons more. Packed with diagrams, patents, and photos, with power shortages now a daily reality in many parts of the world, this book offers a fresh approach very rarely ever mentioned in the mainstream media. Moray B. King is an internationally known physicist and author. He speaks at physics conferences around the world on Zero-Point Energy, and is well known in the world of Tesla Technology, Frontier Sciences and Quantum Physics. ISBN: 0-931882-00-2

204 pages. 6x8 Paperback. Illustrated. $12.95. Code: TAP

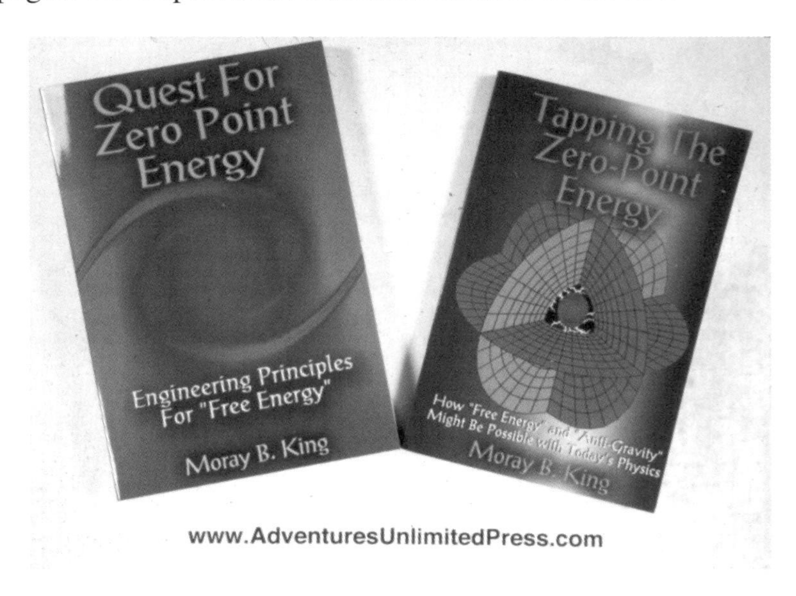

QUEST FOR ZERO-POINT ENERGY
Engineering Principles for "Free Energy"
by Moray B. King

King expands, with diagrams, on how free energy and anti-gravity are possible. The theories of zero point energy maintain there are tremendous fluctuations of electrical field energy embedded within the fabric of space. King explains the following topics: Tapping the Zero-Point Energy as an Energy Source; Fundamentals of a Zero-Point Energy Technology; Vacuum Energy Vortices; The Super Tube; Charge Clusters: The Basis of Zero-Point Energy Inventions; Vortex Filaments, Torsion Fields and the Zero-Point Energy; Transforming the Planet with a Zero-Point Energy Experiment; Dual Vortex Forms: The Key to a Large Zero-Point Energy Coherence. Packed with diagrams, patents and photos. With power shortages now a daily reality in many parts of the world, this book offers a fresh approach very rarely mentioned in the mainstream media.

224 pages. 6x9 Paperback. Illustrated. $14.95. Code: QZPE

ANTI-GRAVITY

THE FREE-ENERGY DEVICE HANDBOOK

A Compilation of Patents and Reports

by David Hatcher Childress

A large-format compilation of various patents, papers, descriptions and diagrams concerning free-energy devices and systems. *The Free-Energy Device Handbook* is a visual tool for experimenters and researchers into magnetic motors and other "over-unity" devices. With chapters on the Adams Motor, the Hans Coler Generator, cold fusion, superconductors, "N" machines, space-energy generators, Nikola Tesla, T. Townsend Brown, and the latest in free-energy devices. Packed with photos, technical diagrams, patents and fascinating information, this book belongs on every science shelf. With energy and profit being a major political reason for fighting various wars, free-energy devices, if ever allowed to be mass distributed to consumers, could change the world! Get your copy now before the Department of Energy bans this book!

292 PAGES. 8X10 PAPERBACK. ILLUSTRATED. BIBLIOGRAPHY. $16.95. CODE: FEH

THE ANTI-GRAVITY HANDBOOK

edited by David Hatcher Childress, with Nikola Tesla, T.B. Paulicki, Bruce Cathie, Albert Einstein and others

The new expanded compilation of material on Anti-Gravity, Free Energy, Flying Saucer Propulsion, UFOs, Suppressed Technology, NASA Cover-ups and more. Highly illustrated with patents, technical illustrations and photos. This revised and expanded edition has more material, including photos of Area 51, Nevada, the government's secret testing facility. This classic on weird science is back in a 90s format!

- **How to build a flying saucer.**
- **Arthur C. Clarke on Anti-Gravity.**
- **Crystals and their role in levitation.**
- **Secret government research and development.**
- **Nikola Tesla on how anti-gravity airships could draw power from the atmosphere.**
- **Bruce Cathie's Anti-Gravity Equation.**
- **NASA, the Moon and Anti-Gravity.**

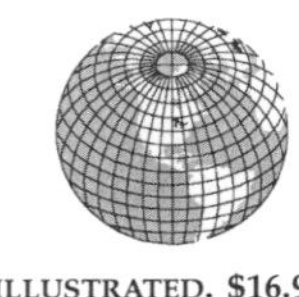

253 PAGES. 7X10 PAPERBACK. BIBLIOGRAPHY/INDEX/APPENDIX. HIGHLY ILLUSTRATED. $16.95. CODE: AGH

ANTI–GRAVITY & THE WORLD GRID

Is the earth surrounded by an intricate electromagnetic grid network offering free energy? This compilation of material on ley lines and world power points contains chapters on the geography, mathematics, and light harmonics of the earth grid. Learn the purpose of ley lines and ancient megalithic structures located on the grid. Discover how the grid made the Philadelphia Experiment possible. Explore the Coral Castle and many other mysteries, including acoustic levitation, Tesla Shields and scalar wave weaponry. Browse through the section on anti-gravity patents, and research resources.

274 PAGES. 7X10 PAPERBACK. ILLUSTRATED. $14.95. CODE: AGW

ANTI–GRAVITY & THE UNIFIED FIELD

edited by David Hatcher Childress

Is Einstein's Unified Field Theory the answer to all of our energy problems? Explored in this compilation of material is how gravity, electricity and magnetism manifest from a unified field around us. Why artificial gravity is possible; secrets of UFO propulsion; free energy; Nikola Tesla and anti-gravity airships of the 20s and 30s; flying saucers as superconducting whirls of plasma; anti-mass generators; vortex propulsion; suppressed technology; government cover-ups; gravitational pulse drive; spacecraft & more.

240 PAGES. 7X10 PAPERBACK. ILLUSTRATED. $14.95. CODE: AGU

ETHER TECHNOLOGY

A Rational Approach to Gravity Control

by Rho Sigma

This classic book on anti-gravity and free energy is back in print and back in stock. Written by a well-known American scientist under the pseudonym of "Rho Sigma," this book delves into international efforts at gravity control and discoid craft propulsion. Before the Quantum Field, there was "Ether." This small, but informative book has chapters on John Searle and "Searle discs;" T. Townsend Brown and his work on anti-gravity and ether-vortex turbines. Includes a forward by former NASA astronaut Edgar Mitchell.

108 PAGES. 6X9 PAPERBACK. ILLUSTRATED. $12.95. CODE: ETT

TAPPING THE ZERO POINT ENERGY

Free Energy & Anti-Gravity in Today's Physics

by Moray B. King

King explains how free energy and anti-gravity are possible. The theories of the zero point energy maintain there are tremendous fluctuations of electrical field energy imbedded within the fabric of space. This book tells how, in the 1930s, inventor T. Henry Moray could produce a fifty kilowatt "free energy" machine; how an electrified plasma vortex creates anti-gravity; how the Pons/Fleischmann "cold fusion" experiment could produce tremendous heat without fusion; and how certain experiments might produce a gravitational anomaly.

190 PAGES. 5X8 PAPERBACK. ILLUSTRATED. $12.95. CODE: TAP

24 hour credit card orders—call: 815-253-6390 fax: 815-253-6300

email: auphq@frontiernet.net www.adventuresunlimitedpress.com www.wexclub.com

FREE ENERGY SYSTEMS

LOST SCIENCE

by Gerry Vassilatos

Rediscover the legendary names of suppressed scientific revolution—remarkable lives, astounding discoveries, and incredible inventions which would have produced a world of wonder. How did the aura research of Baron Karl von Reichenbach prove the vitalistic theory and frighten the greatest minds of Germany? How did the physiophone and wireless of Antonio Meucci predate both Bell and Marconi by decades? How does the earth battery technology of Nathan Stubblefield portend an unsuspected energy revolution? How did the geoaetheric engines of Nikola Tesla threaten the establishment of a fuel-dependent America? The microscopes and virus-destroying ray machines of Dr. Royal Rife provided the solution for every world-threatening disease. Why did the FDA and AMA together condemn this great man to Federal Prison? The static crashes on telephone lines enabled Dr. T. Henry Moray to discover the reality of radiant space energy. Was the mysterious "Swedish stone," the powerful mineral which Dr. Moray discovered, the very first historical instance in which stellar power was recognized and secured on earth? Why did the Air Force initially fund the gravitational warp research and warp-cloaking devices of T. Townsend Brown and then reject it? When the controlled fusion devices of Philo Farnsworth achieved the "break-even" point in 1967 the FUSOR project was abruptly cancelled by ITT.

304 PAGES. 6X9 PAPERBACK. ILLUSTRATED. BIBLIOGRAPHY. $16.95. CODE: LOS

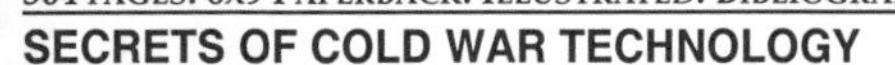

SECRETS OF COLD WAR TECHNOLOGY

Project HAARP and Beyond

by Gerry Vassilatos

Vassilatos reveals that "Death Ray" technology has been secretly researched and developed since the turn of the century. Included are chapters on such inventors and their devices as H.C. Vion, the developer of auroral energy receivers; Dr. Selim Lemstrom's pre-Tesla experiments; the early beam weapons of Grindell-Mathews, Ulivi, Turpain and others; John Hettenger and his early beam power systems. Learn about Project Argus, Project Teak and Project Orange; EMP experiments in the 60s; why the Air Force directed the construction of a huge Ionospheric "backscatter" telemetry system across the Pacific just after WWII; why Raytheon has collected every patent relevant to HAARP over the past few years; more.

250 PAGES. 6X9 PAPERBACK. ILLUSTRATED. $15.95. CODE: SCWT

QUEST FOR ZERO-POINT ENERGY

Engineering Principles for "Free Energy"

by Moray B. King

King expands, with diagrams, on how free energy and anti-gravity are possible. The theories of zero point energy maintain there are tremendous fluctuations of electrical field energy embedded within the fabric of space. King explains the following topics: Tapping the Zero-Point Energy as an Energy Source; Fundamentals of a Zero-Point Energy Technology; Vacuum Energy Vortices; The Super Tube; Charge Clusters: The Basis of Zero-Point Energy Inventions; Vortex Filaments, Torsion Fields and the Zero-Point Energy; Transforming the Planet with a Zero-Point Energy Experiment; Dual Vortex Forms: The Key to a Large Zero-Point Energy Coherence. Packed with diagrams, patents and photos. With power shortages now a daily reality in many parts of the world, this book offers a fresh approach very rarely mentioned in the mainstream media.

224 PAGES. 6X9 PAPERBACK. ILLUSTRATED. $14.95. CODE: QZPE

THE TIME TRAVEL HANDBOOK

A Manual of Practical Teleportation & Time Travel

edited by David Hatcher Childress

In the tradition of *The Anti-Gravity Handbook* and *The Free-Energy Device Handbook*, science and UFO author David Hatcher Childress takes us into the weird world of time travel and teleportation. Not just a whacked-out look at science fiction, this book is an authoritative chronicling of real-life time travel experiments, teleportation devices and more. *The Time Travel Handbook* takes the reader beyond the government experiments and deep into the uncharted territory of early time travellers such as Nikola Tesla and Guglielmo Marconi and their alleged time travel experiments, as well as the Wilson Brothers of EMI and their connection to the Philadelphia Experiment—the U.S. Navy's forays into invisibility, time travel, and teleportation. Childress looks into the claims of time travelling individuals, and investigates the unusual claim that the pyramids on Mars were built in the future and sent back in time. A highly visual, large format book, with patents, photos and schematics. Be the first on your block to build your own time travel device!

316 PAGES. 7X10 PAPERBACK. ILLUSTRATED. $16.95. CODE: TTH

THE TESLA PAPERS

Nikola Tesla on Free Energy & Wireless Transmission of Power

by Nikola Tesla, edited by David Hatcher Childress

David Hatcher Childress takes us into the incredible world of Nikola Tesla and his amazing inventions. Tesla's rare article "The Problem of Increasing Human Energy with Special Reference to the Harnessing of the Sun's Energy" is included. This lengthy article was originally published in the June 1900 issue of *The Century Illustrated Monthly Magazine* and it was the outline for Tesla's master blueprint for the world. Tesla's fantastic vision of the future, including wireless power, anti-gravity, free energy and highly advanced solar power. Also included are some of the papers, patents and material collected on Tesla at the Colorado Springs Tesla Symposiums, including papers on: •The Secret History of Wireless Transmission •Tesla and the Magnifying Transmitter •Design and Construction of a Half-Wave Tesla Coil •Electrostatics: A Key to Free Energy •Progress in Zero-Point Energy Research •Electromagnetic Energy from Antennas to Atoms •Tesla's Particle Beam Technology •Fundamental Excitatory Modes of the Earth-Ionosphere Cavity

325 PAGES. 8X10 PAPERBACK. ILLUSTRATED. $16.95. CODE: TTP

THE FANTASTIC INVENTIONS OF NIKOLA TESLA

by Nikola Tesla with additional material by David Hatcher Childress

This book is a readable compendium of patents, diagrams, photos and explanations of the many incredible inventions of the originator of the modern era of electrification. In Tesla's own words are such topics as wireless transmission of power, death rays, and radio-controlled airships. In addition, rare material on German bases in Antarctica and South America, and a secret city built at a remote jungle site in South America by one of Tesla's students, Guglielmo Marconi. Marconi's secret group claims to have built flying saucers in the 1940s and to have gone to Mars in the early 1950s! Incredible photos of these Tesla craft are included. The Ancient Atlantean system of broadcasting energy through a grid system of obelisks and pyramids is discussed, and a fascinating concept comes out of one chapter: that Egyptian engineers had to wear protective metal head-shields while in these power plants, hence the Egyptian Pharoah's head covering as well as the Face on Mars! •His plan to transmit free electricity into the atmosphere. •How electrical devices would work using only small antennas. •Why unlimited power could be utilized anywhere on earth. •How radio and radar technology can be used as death-ray weapons in Star Wars.

342 PAGES. 6X9 PAPERBACK. ILLUSTRATED. $16.95. CODE: FINT

HISTORY—CONSPIRACY

POPULAR PARANOIA

The Best of Steamshovel Press

edited by Kenn Thomas

The anthology exposes the biologocal warfare origins of AIDS; the Nazi/Nation of Islam link; the cult of Elizabeth Clare Prophet; the Oklahoma City bombing writings of the late Jim Keith, as well as an article on Keith's own strange death; the conspiratorial mind of John Judge; Marion Pettie and the shadowy Finders group in Washington, DC; demonic iconography; the death of Princess Diana, its connection to the Octopus and the Saudi aerospace contracts; spies among the Rajneeshis; scholarship on the historic Illuminati; and many other parapolitical topics. The book also includes the Steamshovel's last-ever interviews with the great Beat writers Allen Ginsberg and William S. Burroughs, and neuronaut Timothy Leary, and new views of the master Beat, Neal Cassady and Jack Kerouac's science fiction.

308 PAGES. 8X10 PAPERBACK. ILLUSTRATED. $19.95. CODE: POPA

THE ORION PROPHECY

Egyptian and Mayan Prophecies on the Cataclysm of 2012

by Patrick Geryl and Gino Ratinckx

In the year 2012 the Earth awaits a super catastrophe: its magnetic field will reverse in one go. Phenomenal earthquakes and tidal waves will completely destroy our civilization. Europe and North America will shift thousands of kilometers northwards into polar climes. Nearly everyone will perish in the apocalyptic happenings. These dire predictions stem from the Mayans and Egyptians—descendants of the legendary Atlantis. The Atlanteans had highly evolved astronomical knowledge and were able to exactly predict the previous world-wide flood in 9792 BC. They built tens of thousands of boats and escaped to South America and Egypt. In the year 2012 Venus, Orion and several others stars will take the same 'code-positions' as in 9792 BC! For thousands of years historical sources have told of a forgotten time capsule of ancient wisdom located in a labyrinth of secret chambers filled with artifacts and documents from the previous flood. We desperately need this information now—and this book gives one possible location.

324 PAGES. 6X9 PAPERBACK. ILLUSTRATED. BIBLIOGRAPHY. $16.95. CODE: ORP

THE SHADOW GOVERNMENT

9-11 and State Terror

by Len Bracken, introduction by Kenn Thomas

Bracken presents the alarming yet convincing theory that nation-states engage in or allow terror to be visited upon their citizens. It is not just liberation movements and radical groups that deploy terroristic tactics for offensive ends. States use terror defensively to directly intimidate their citizens and to indirectly attack themselves or harm their citizens under a false flag. Their motives? To provide pretexts for war or for increased police powers or both. This stratagem of indirectly using terrorism has been executed by statesmen in various ways but tends to involve the pretense of blind eyes, misdirection, and cover-ups that give statesmen plausible deniability. Lusitiania, Pearl Harbor, October Surprise, the first World Trade Center bombing, the Oklahoma City bombing and other well-known incidents suggest that terrorism is often and successfully used by states in an indirectly defensive way to take the offensive against enemies at home and abroad. Was 9-11 such an indirect defensive attack?

288 PAGES. 6X9 PAPERBACK. ILLUSTRATED. $16.00. CODE: SGOV

MASS CONTROL

Engineering Human Consciousness

by Jim Keith

Conspiracy expert Keith's final book on mind control, Project Monarch, and mass manipulation presents chilling evidence that we are indeed spinning a Matrix. Keith describes the New Man, whose conception of reality is a dance of electronic images fired into his forebrain, a gossamer construction of his masters, designed so that he will not—under any circumstances—perceive the actual. His happiness is delivered to him through a tube or an electronic connection. His God lurks behind an electronic curtain; when the curtain is pulled away we find the CIA sorcerer, the media manipulator... Chapters on the CIA, Tavistock, Jolly West and the Violence Center, Guerrilla Mindwar, Brice Taylor, other recent "victims," more.

256 PAGES. 6X9 PAPERBACK. ILLUSTRATED. INDEX. $16.95. CODE: MASC

WAKE UP DOWN THERE!

The Excluded Middle Anthology

by Greg Bishop

The great American tradition of dropout culture makes it over the millennium mark with a collection of the best from *The Excluded Middle*, the critically acclaimed underground zine of UFOs, the paranormal, conspiracies, psychedelia, and spirit. Contributions from Robert Anton Wilson, Ivan Stang, Martin Kottmeyer, John Shirley, Scott Corrales, Adam Gorightly and Robert Sterling; and interviews with James Moseley, Karla Turner, Bill Moore, Kenn Thomas, Richard Boylan, Dean Radin, Joe McMoneagle, and the mysterious Ira Einhorn (an *Excluded Middle* exclusive). Includes full versions of interviews and extra material not found in the newsstand versions.

420 PAGES. 8X11 PAPERBACK. ILLUSTRATED. $25.00. CODE: WUDT

DARK MOON

Apollo and the Whistleblowers

by Mary Bennett and David Percy

- Was Neil Armstrong really the first man on the Moon?
- Did you know a second craft was going to the Moon at the same time as Apollo 11?
- Do you know that potentially lethal radiation is prevalent throughout deep space?
- Do you know there are serious discrepancies in the account of the Apollo 13 'accident'?
- Did you know that 'live' color TV from the Moon was not actually live at all?
- Did you know that the Lunar Surface Camera had no viewfinder?
- Do you know that lighting was used in the Apollo photographs—yet no lighting equipment was taken to the Moon?

All these questions, and more, are discussed in great detail by British researchers Bennett and Percy in *Dark Moon*, the definitive book (nearly 600 pages) on the possible faking of the Apollo Moon missions. Bennett and Percy delve into every possible aspect of this beguiling theory, one that rocks the very foundation of our beliefs concerning NASA and the space program. Tons of NASA photos analyzed for possible deceptions.

568 PAGES. 6X9 PAPERBACK. ILLUSTRATED. BIBLIOGRAPHY. INDEX. $25.00. CODE: DMO

MYSTIC TRAVELLER SERIES

THE MYSTERY OF EASTER ISLAND
by Katherine Routledge

The reprint of Katherine Routledge's classic archaeology book which was first published in London in 1919. The book details her journey by yacht from England to South America, around Patagonia to Chile and on to Easter Island. Routledge explored the amazing island and produced one of the first-ever accounts of the life, history and legends of this strange and remote place. Routledge discusses the statues, pyramid-platforms, Rongo Rongo script, the Bird Cult, the war between the Short Ears and the Long Ears, the secret caves, ancient roads on the island, and more. This rare book serves as a sourcebook on the early discoveries and theories on Easter Island.

432 PAGES. 6X9 PAPERBACK. ILLUSTRATED. $16.95. CODE: MEI

MYSTERY CITIES OF THE MAYA
Exploration and Adventure in Lubaantun & Belize
by Thomas Gann

First published in 1925, *Mystery Cities of the Maya* is a classic in Central American archaeology-adventure. Gann was close friends with Mike Mitchell-Hedges, the British adventurer who discovered the famous crystal skull with his adopted daughter Sammy and Lady Richmond Brown, their benefactress. Gann battles pirates along Belize's coast and goes upriver with Mitchell-Hedges to the site of Lubaantun where they excavate a strange lost city where the crystal skull was discovered. Lubaantun is a unique city in the Mayan world as it is built out of precisely carved blocks of stone without the usual plaster-cement facing. Lubaantun contained several large pyramids partially destroyed by earthquakes and a large amount of artifacts. Gann shared Mitchell-Hedges belief in Atlantis and lost civilizations (pre-Mayan) in Central America and the Caribbean. Lots of good photos, maps and diagrams.

252 PAGES. 6X9 PAPERBACK. ILLUSTRATED. $16.95. CODE: MCOM

IN SECRET TIBET
by Theodore Illion

Reprint of a rare 30s adventure travel book. Illion was a German wayfarer who not only spoke fluent Tibetan, but travelled in disguise as a native through forbidden Tibet when it was off-limits to all outsiders. His incredible adventures make this one of the most exciting travel books ever published. Includes illustrations of Tibetan monks levitating stones by acoustics.

210 PAGES. 6X9 PAPERBACK. ILLUSTRATED. $15.95. CODE: IST

DARKNESS OVER TIBET
by Theodore Illion

In this second reprint of Illion's rare books, the German traveller continues his journey through Tibet and is given directions to a strange underground city. As the original publisher's remarks said, "this is a rare account of an underground city in Tibet by the only Westerner ever to enter it and escape alive! "

210 PAGES. 6X9 PAPERBACK. ILLUSTRATED. $15.95. CODE: DOT

DANGER MY ALLY
The Amazing Life Story of the Discoverer of the Crystal Skull
by "Mike" Mitchell-Hedges

The incredible life story of "Mike" Mitchell-Hedges, the British adventurer who discovered the Crystal Skull in the lost Mayan city of Lubaantun in Belize. Mitchell-Hedges has lived an exciting life: gambling everything on a trip to the Americas as a young man, riding with Pancho Villa, questing for Atlantis, fighting bandits in the Caribbean and discovering the famous Crystal Skull.

374 PAGES. 6X9 PAPERBACK. ILLUSTRATED. BIBLIOGRAPHY & INDEX. $16.95. CODE: DMA

IN SECRET MONGOLIA
by Henning Haslund

First published by Kegan Paul of London in 1934, Haslund takes us into the barely known world of Mongolia of 1921, a land of god-kings, bandits, vast mountain wilderness and a Russian army running amok. Starting in Peking, Haslund journeys to Mongolia as part of the Krebs Expedition—a mission to establish a Danish butter farm in a remote corner of northern Mongolia. Along the way, he smuggles guns and nitroglycerin, is thrown into a prison by the new Communist regime, battles the Robber Princess and more. With Haslund we meet the "Mad Baron" Ungern-Sternberg and his renegade Russian army, the many characters of Urga's fledgling foreign community, and the last god-king of Mongolia, Seng Chen Gegen, the fifth reincarnation of the Tiger god and the "ruler of all Torguts." Aside from the esoteric and mystical material, there is plenty of just plain adventure: Haslund encounters a Mongolian werewolf; is ambushed along the trail; escapes from prison and fights terrifying blizzards; more.

374 PAGES. 6X9 PAPERBACK. ILLUSTRATED. BIBLIOGRAPHY & INDEX. $16.95. CODE: ISM

MEN & GODS IN MONGOLIA
by Henning Haslund

First published in 1935 by Kegan Paul of London, Haslund takes us to the lost city of Karakota in the Gobi desert. We meet the Bodgo Gegen, a god-king in Mongolia similar to the Dalai Lama of Tibet. We meet Dambin Jansang, the dreaded warlord of the "Black Gobi." There is even material in this incredible book on the Hi-mori, an "airhorse" that flies through the sky (similar to a Vimana) and carries with it the sacred stone of Chintamani. Aside from the esoteric and mystical material, there is plenty of just plain adventure: Haslund and companions journey across the Gobi desert by camel caravan; are kidnapped and held for ransom; withness initiation into Shamanic societies; meet reincarnated warlords; and experience the violent birth of "modern" Mongolia.

358 PAGES. 6X9 PAPERBACK. ILLUSTRATED. INDEX. $15.95. CODE: MGM

One Adventure Place
P.O. Box 74
Kempton, Illinois 60946
United States of America
•Tel.: 1-800-718-4514 or 815-253-6390
•Fax: 815-253-6300
Email: auphq@frontiernet.net
http://www.adventuresunlimitedpress.com
or www.adventuresunlimited.nl

10% Discount when you order 3 or more items!

ORDERING INSTRUCTIONS

- ✓ Remit by USD$ Check, Money Order or Credit Card
- ✓ Visa, Master Card, Discover & AmEx Accepted
- ✓ Prices May Change Without Notice
- ✓ 10% Discount for 3 or more Items

SHIPPING CHARGES

United States

- ✓ Postal Book Rate { $3.00 First Item / 50¢ Each Additional Item
- ✓ Priority Mail { $4.50 First Item / $2.00 Each Additional Item
- ✓ UPS { $5.00 First Item / $1.50 Each Additional Item

NOTE: UPS Delivery Available to Mainland USA Only

Canada

- ✓ Postal Book Rate { $6.00 First Item / $2.00 Each Additional Item
- ✓ Postal Air Mail { $8.00 First Item / $2.50 Each Additional Item
- ✓ Personal Checks or Bank Drafts MUST BE USD$ and Drawn on a US Bank
- ✓ Canadian Postal Money Orders in US$ OK
- ✓ Payment MUST BE US$

All Other Countries

- ✓ Surface Delivery { $10.00 First Item / $4.00 Each Additional Item
- ✓ Postal Air Mail { $14.00 First Item / $5.00 Each Additional Item
- ✓ Checks and Money Orders MUST BE US$ and Drawn on a US Bank or branch.
- ✓ Payment by credit card preferred!

SPECIAL NOTES

- ✓ RETAILERS: Standard Discounts Available
- ✓ BACKORDERS: We Backorder all Out-of-Stock Items Unless Otherwise Requested
- ✓ PRO FORMA INVOICES: Available on Request
- ✓ VIDEOS: NTSC Mode Only. Replacement only.
- ✓ For PAL mode videos contact our other offices:

European Office:
Adventures Unlimited, Pannewal 22, Enkhuizen, 1602 KS, The Netherlands
http: www.adventuresunlimited.nl
Check Us Out Online at:
www.adventuresunlimitedpress.com

Please check: ☑

❐ This is my first order ❐ I have ordered before ❐ This is a new address

Name	
Address	
City	
State/Province	Postal Code
Country	
Phone day	Evening
Fax	Email

Item Code	Item Description	Price	Qty	Total

Please check: ☑

- ❐ Postal-Surface
- ❐ Postal-Air Mail (Priority in USA)
- ❐ UPS (Mainland USA only)

Subtotal ➡	
Less Discount-10% for 3 or more items ➡	
Balance ➡	
Illinois Residents 6.25% Sales Tax ➡	
Previous Credit ➡	
Shipping ➡	
Total (check/MO in USD$ only) ➡	

❐ Visa/MasterCard/Discover/Amex

Card Number

Expiration Date

10% Discount When You Order 3 or More Items!

Comments & Suggestions

Share Our Catalog with a Friend